The Physics of Low Dimensional Materials

The Physics of Low Dimensional Materials

Frank J Owens

City University of New York, USA

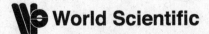

World Scientific

NEW JERSEY · LONDON · SINGAPORE · BEIJING · SHANGHAI · HONG KONG · TAIPEI · CHENNAI · TOKYO

Published by

World Scientific Publishing Co. Pte. Ltd.

5 Toh Tuck Link, Singapore 596224

USA office: 27 Warren Street, Suite 401-402, Hackensack, NJ 07601

UK office: 57 Shelton Street, Covent Garden, London WC2H 9HE

British Library Cataloguing-in-Publication Data
A catalogue record for this book is available from the British Library.

THE PHYSICS OF LOW DIMENSIONAL MATERIALS

ISBN 978-981-3225-85-5

Desk Editor: Christopher Teo

Typeset by Stallion Press
Email: enquiries@stallionpress.com

Printed in Singapore

Preface

For many years solid state physicists believed that two dimensional solids are unstable. However, that prediction was brought into question with the fabrication of graphene in 2003. Graphene is a two dimensional solid of carbon atoms having the same arrangement as in the planes of graphite. The relatively simple fabrication and the unique electronic properties, such as the conduction electrons behaving like relativistic Fermions, motivated intense research activity. The possibility of faster switching field effect transistors was envisioned. The fact that graphene had no band gap at certain points in the Brillouin zone making it a semimetal was considered an obstacle to developing it into electronic devices. Field effect transistors are typically made of silicon which has a gap at all points in the zone. Research efforts on graphene turned to examining the properties of graphene ribbons or doped graphene. Recently researchers began to explore the possibility of other two dimensional materials, such as black phosphorous and dichalcognides, which could be fabricated by exfoliation of layered materials or other methods.

Of course before the discovery of graphene, single walled carbon and boron nitride were synthesized — they are effectively one dimensional materials. There also exist a number of crystals which consist of weakly interacting chains such as Se which effectively make these crystals containers for one dimensional materials. While isolation of a single chain from such solids has not been achieved it is of interest to consider their properties.

The purpose of this book is two-fold. First to explain the properties of low dimensional solids such as electronic, vibrational and magnetic structure in terms of simple models, that have been used to account for the properties of three dimensional materials providing an elementary introduction to the physics of low dimensional materials. The second objective is to examine the properties of other possible low dimensional materials that are now the subject of research.

The first chapter deals with theoretical models such as density functional theory which is used to obtain the properties of solids. This is done because theoretical models will be used in subsequent chapters to obtain the fundamental properties of the newer low dimensional materials where experimental results are unavailable. Chapters 2 and 3 use simple models of the solid state to explain the electronic and vibrational structure of two dimensional materials. Chapters 5, 6, and 7 present the properties of many different kinds of low dimensional materials. Chapter 8 is concerned with the magnetic properties of these materials.

Chapter 9 discusses the properties of superconductivity in low dimensional materials. Throughout the book various experimental methods used to measure properties are described and results of their measurements presented.

About the Author

Frank J Owens Ph. D is a research professor in the department of physics of Hunter College of the City University of New York. Previously he has held positions as a senior research scientist at the Army Armament Research and Development and Engineering Center, the graduate faculty of the City University of New York, the National Aeronautics and Space Administration, Goddard Space Flight Center, New Jersey Institute of Technology, and Farliegh Dickenson University. He is the author or co-author of 6 books on nanotechnology and superconductivity. He was selected as a secretary of the Army fellow and is a fellow of the American physical society.

Contents

Chapter 1

Computational Material Science

1.1 Introduction

The field of computational material science has grown in the last thirty years to become an important part of materials research. There have been numerous applications to low dimensional materials. The subsequent chapters will use the results of such calculations to describe the vibrational and electronic properties of low dimensional materials. Thus the subject is discussed in the first chapter. Computational materials science refers to theoretically predicting the properties of materials such as the structure, vibrational and electronic properties. It can and has been used to predict the existence of new materials. For example the possibility of the existence of C_{60} and its properties such as the vibrational frequencies were predicted long before its discovery. Theoretical modeling can also be used to determine the properties of existing materials and is particularly usefull in predicting properties which are difficult to determine experimentally. Calculations of elastic constants give results quite close to experimental values and are much easier to do compared to experimental methods. The development of computational material science as an important tool in materials research is largely a result of the development of high performance computers and theoretical models such as Density Functional Theory (DFT). A critical issue in modeling solids is choice of the size of the solids, meaning the number of atoms or ions in the system. The number of substituents treated must be such that the results give values of properties corresponding to macroscopic systems. A cubic micron of copper has approximately 10^4 atoms which is near the limit of numbers that can be handled by present day computers using the most sophisticated theoretical models. Another limitation is the time scale which should be less than 10^{-15} seconds in order to deal with atomic vibrations in the materials.

Theoretical modeling has been applied to carbon nanotubes and graphene. Perhaps because of the smaller size of these nanostructures which typically have less than 10^6 atoms these limitations may be less of a problem. A particular objective of this work is to predict new and interesting modifications of low dimensional materials which have application potential. This can provide guidance for identifying interesting materials to synthesize. For example boron nitride nanotubes, which have many advantages compared to carbon nanotubes, have band gaps greater than 5 eV limiting their electronic application potential. However, theoretical modeling has predicted that increasing the boron content relative to the nitrogen content or applying an electric field can significantly reduce the band gap to a value that makes them semiconductors. This may allow for the possibility of electronic applications.

In this chapter an overview of the theoretical approaches used to predict the properties of low dimensional materials will be presented as well as some examples of calculations and their predictions.

1.2 *Ab Initio* Molecular Orbital Theory

Various modifications of molecular orbital theory have been used to calculate the electronic and vibrational properties as well as geometric conformation of solids. The object of molecular orbital theory is to obtain approximate solutions of the non relativistic Schrödinger wave equation for a system of many atoms and electrons, which is given by,

$$\left[-\sum (h/\pi)^2 / 2m(\nabla_i^2) + \sum_{i<j} e^2/r_{ij} - \sum_{iA} Z_A e^2/r_{Ai} \right.$$

$$\left. + \sum_{A<B} Z_A Z_B e^2/R_{AB} \right] \psi_e = E\psi_e \qquad (1.1)$$

The first term is the kinetic energy of the electrons where,

$$\nabla_i^2 = d^2/dx_i^2 + d^2/dY_i^2 + d^2/dz_i^2 \qquad (1.2)$$

The second term represents the electrostatic repulsion between the electrons. The third term is the electrostatic attraction between the electrons and the nuclei and the fourth term is the electrostatic repulsion between the nuclei. E is the energy and ψ_e is the wave function. Equation (1.1) omits the kinetic energy of the nuclei because the electrons move much more rapidly than the nuclei which are much heavier. In effect the motion of the nuclei is ignored this is referred to as the Born–Oppenheimer approximation. The first successful solution of the Schrödinger equation of the hydrogen molecule was by Heitler and London in 1927.[1] In order to get some understanding of how the problem is solved let us consider the simplest of all molecules, the H_2 plus molecular ion. Referring to Figure 1.1 the Hamiltonian

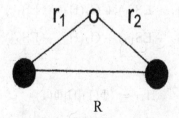

Figure 1.1 Structure of H_2^+ molecular ion.

of the molecule is,

$$H = H_1 + H_2 + e/R \qquad (1.3)$$

e/R is the electrostatic repulsion between the two nuclei. H_1 and H_2 are the Hamiltonians of the separated hydrogen atoms given by,

$$H_1 = -(h/\pi)^2/2m(\nabla_i^2) - e/r_1 \qquad (1.4)$$

and,

$$H_2 = -(h/\pi)^2/2m(\nabla_2^2) - e/r_2 \qquad (1.5)$$

The object is to find the ground state energy and the wave function.

A trial wave function is constructed which is a linear combination of the 1s orbital of each hydrogen atom, i.e.

$$\psi = C_1 \Phi(1) + C_2 \Phi(2) \qquad (1.6)$$

The variation principle can be used where the constants C_1 and C_2 can be varied to find the minimum energy i.e.

$$dE/dC_1 = 0 \quad \text{and} \quad dE/dC_2 = 0. \qquad (1.7)$$

The energy E is given by,[*]

$$E = \langle C_1 \Phi(1) + C_2 \Phi(2)|H|C_1 \Phi(1) + C_2 \Phi(2)\rangle /$$
$$\langle C_1 \Phi(1) + C_2 \Phi(2)|C_1 \Phi(1) + C_2 \Phi(2)\rangle. \qquad (1.8)$$

The application of equations (1.7) to E yields two simultaneous equations.

$$C_1(H_{11} - ES_{11}) + C_2(H_{12} - ES_{12}) = 0 \qquad (1.9)$$

$$C_1(H_{12} - ES_{12}) + C_2(H_{22} - ES_{22}) = 0 \qquad (1.10)$$

where

$$H_{11} = \langle \Phi(1)|H|\Phi(1)\rangle \qquad (1.11)$$

$$H_{12} = \langle \Phi(1)|H|\Phi(2)\rangle \qquad (1.12)$$

---

[*]The notation $<\ >$ means the volume integral for example $< \psi\psi* >$ is $\int \psi\psi*dV$

$$S_{11} = \langle \Phi(1)|\Phi(1) \rangle \qquad (1.13)$$

$$S_{12} = \langle \Phi(1)|\Phi(2) \rangle \qquad (1.14)$$

The solution of equations (1.9) and (1.10) yields two energy levels given by,

$$E = [H_{11} + H_{12}]/[1 + S_{12}] \qquad (1.15)$$

$$E = [H_{11} - H_{12}]/[1 - S_{12}] \qquad (1.16)$$

This means that when two hydrogen atoms bond to form a H_2^+ molecular ion, the 1s level of the H atom is split into two levels referred to as a bonding and antibonding level. Figure 1.2 illustrates the energy levels of the hydrogen molecular ion. The corresponding bonding and antibonding wave functions are given by,

$$\psi_B = 1/[2^{1/2}(1 + S_{12})][\Phi(1) + \Phi(2)] \qquad (1.17)$$

$$\psi_A = 1/[2^{1/2}(1 - S_{12})][\Phi(1) - \Phi(2)]. \qquad (1.18)$$

Figure 1.3 shows a plot of the electron probability density in a plane containing the two H atoms for these two orbitals i.e. ψ_B^2 and ψ_A^2. It is seen that the bonding orbital has its density concentrated between the two nuclei whereas the antibonding orbital does not. Even with

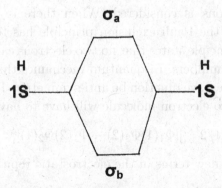

Figure 1.2 Calculated ground state energy levels of H_2^+ molecular ion showing a splitting of the hydrogen 1s orbitals into bonding and antibonding orbitals.

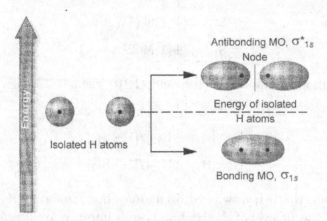

Figure 1.3 Map of the calculated electron density of the H_2^+ molecular ion for the bonding and antibonding orbitals in a plane containing the two H atoms.

this simplest of molecules one can see some of the calculational difficulties such as the need to evaluate integrals such as $\langle \Phi(1)|\Phi(2)\rangle$ referred to as overlap integrals.

The H_2^+ molecule has only one electron and thus electron-electron repulsion is not present. If the neutral H_2 molecule were to be considered, the electrostatic repulsion between the two electrons would have to be included which would have the form,

$$\langle \Phi(1)[1/r_{12}]\Phi(2)\rangle. \tag{1.19}$$

The calculational obstacles become very large when a molecule having many electrons is considered. When there is more than one electron present the Pauli exclusion principle has to be taken into account. The principle states that no two electrons can have the same set of quantum numbers. In quantum mechanics this is handled by requiring that the wave function be anti-symmetric. Thus a trial wave function for a two electron molecule will have to have the form,

$$[1/2^{1/2}][\Phi_1(1)\Phi_2(2) - \Phi_1(2)\Phi_2(1)]. \tag{1.20}$$

This will lead to new terms in the electrostatic repulsion having the form,

$$\langle \Phi_1(1)\Phi_2(2)[1/r_{12}]\Phi_1(2)\Phi_2(1)\rangle. \tag{1.21}$$

This term is called the exchange energy and represents the difference in the electrostatic repulsion between two electrons that have their spins parallel and antiparallel.

1.3 Density Functional Theory

Density functional theory is an extension of the concept of the Thomas–Fermi model of a many electron atom to molecules and solids. The theory treats the electrons as a gas of free electrons confined to a volume V about the nucleus by a spherically symmetric potential. In the context of the free electron model of metals for N electrons confined to a cubic volume V, the Fermi level, the highest filled level for a three dimensional solid is,

$$E_f = [(h/\pi)^2/2m][3\pi^2 N/V]^{2/3}. \tag{1.22}$$

The depth of the potential at any value of r can be related to the density of electrons for that value of r by assuming that the depth of the potential is such that the energy levels are filled to the top i.e. $E_f = -V[r]$. This yields a relationship between the potential and the density of electrons, $\rho = N/V$ given by,

$$-V[r] = [(h/\pi)^2/2m][3\pi^2 \rho]^{2/3}. \tag{1.23}$$

The model assumes the V(r) does not change significantly in lengths compared to the wave lengths of the electrons. This means a number of electrons can be localized in a volume in which V(r) is essentially constant. The importance of this result is that it transforms the electrostatic interaction of the electrons with the nuclei and the other electrons (the second and third terms in equation (1.1) to a form where the interaction between each electron and every other electron can be represented by the interaction of the electron and nuclei with a charge density. For a molecule having many nuclei and electrons, the electrostatic interaction of the electrons with each other and the nucleus becomes that of non interacting electrons in an effective potential which is a function of the charge density. The Thomas –Fermi model of the electronic structure of atoms did not predict the energy levels of atoms very well primarily because it did not include the exchange interaction. However, it did provide a conceptual basis for the development of DFT by proposing that the electron repulsive interaction between the electrons could be replaced by a charge density.

In DFT, the ground state energy of a many electron system is a function only of the electron density $\rho[r]$. The wave function must satisfy a Schrödinger-like equation having the form,

$$\left[-\sum (h/\pi)^2/2m(\nabla_i^2) + V_N[r] + \int \rho[r']/|\mathbf{r} - \mathbf{r}'|d^3r' + \epsilon(\rho[r]) \right] \psi_i(\mathbf{r})$$

$$= E_i\psi_i(\mathbf{r}) \tag{1.24}$$

All terms except the electrostatic repulsion between the nuclei are a function of the charge density. The first term is the kinetic energy. The kinetic energy density at each point is assumed to correspond to the kinetic energy density of a homogeneous non interacting electron gas which has been shown to be proportional to $\rho^{5/3}$. The second term is the interaction of the electrons with the nucleus. The third term is the electrostatic repulsion of the electrons with each other in terms of an electron density. The last terms represents the exchange and correlation interactions. In DFT accurate formulas for ϵ have been developed from simulations of a uniform non-interacting electron gas. The model used to calculate the exchange interaction in DFT is based on the idea to treat the electronic structure of a simple metal such as lithium as a homogenous gas of electrons around a lattice of positive charges. This model is used in DFT because it is the only one which yields an exact and accurate form of the exchange interaction. The model is referred to as the local density approximation (LDA) and the exchange term has the form,

$$E_{xc}^{LDA}[\rho] = \int \rho[r]\epsilon_{xc}(\rho[r])dr \tag{1.25}$$

where $\epsilon_{xc}(\rho[r])$ is the energy density, i.e., the exchange plus correlation energy per electron. The use of the variational method to find the energy of a many electron system always leads to a larger energy than the exact energy. The difference between the two energies is called the correlation energy. In a homogeneous electron gas with electron density "ρ", it is assumed that the electron density "ρ" varies slowly with position. The specific form of the exchange energy was obtained by Slater and is given by,

$$\epsilon_{xc} = (-3/4)[3\rho[r]/\pi]^{1/3} \tag{1.26}$$

Substituting equation (1.26) into (1.25) gives the following form for the exchange energy,

$$E_{xc}^{LDA}[\rho] = (-3/4)[3/\pi]^{1/3} \int \rho[r]^{4/3} dr \tag{1.27}$$

A brief over view of DFT approach within the LDA approximation has been presented without going into the details and proofs of the Hohenberg-Kohn existence and variational theorems. The basic understanding that the reader should be left with is that DFT transforms the complex many body problem of interacting electrons in the external potential of the nucleus to a tractable problem of non interacting electrons moving in an effective potential which is a function of the electron charge density. With the DFT approach it became possible to perform calculations on quite large molecules with good to excellent accuracies.

1.4 Some Examples of the Application of DFT

1.4.1 *Properties of C_{60}*

Using the example of C_{60} this section will examine how well DFT predicts the properties of this structure.[3] Figure 1.4 shows the structure of C_{60}. It consists of 12 pentagonal (5 sided) and 20 hexagonal (6 sided) carbon rings symmetrically arranged to form a soccer ball like structure. Table I compares the bonds lengths of the double and single carbon bonds of C_{60} calculated by density functional theory (DFT) with experimental values. DFT method predicts bond lengths in good agreement with experiment. The most intense observed Raman vibrational mode of C_{60} is the pentagonal pinch mode which involves an expansion and contraction of the carbon pentagonal ring

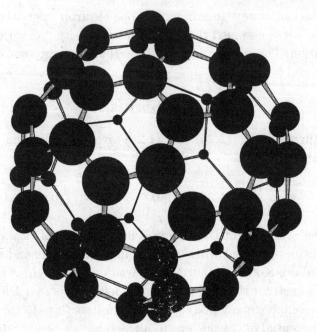

Figure 1.4   Structure of the C_{60} molecule calculated by density functional theory. (Ref. 3)

Table I   Comparison of some experimental and calculated properties of C_{60} using DFT.

Property	Experimental	DFT
C-C (A)	1.46	1.46
C=C (A)	1.40	1.40
IP (eV)	7.6 ± 0.2	7.22
ω(cm-1)	1493	1455

DFT calculations were performed using B3LYP/ 6-31G, Ref. 3

parallel to the surface of the ball. The DFT calculation predicts the frequency of this mode to within 2.5% of the experimental value. In order to examine the ability of the models to predict electronic structure, the results of calculating the vertical ionization energy are shown. This is calculated by taking the difference in the total energy of the minimum energy structure of C_{60} and the ion C_{60}^+ having the same structure. Again the DFT model gives reasonable agreement with the experimental value. The results such as these and the need for less computer time compared to ab initio methods are the reasons that DFT theory is widely used to predict the properties of materials. It is also possible to employ periodic boundary conditions using DFT which enables a calculation of the energy levels of solids as a function of the wave vector. The concept of periodic boundary conditions will be discussed in Chapter 2. Throughout the volume DFT will be used to obtain the energy levels and vibrational properties of low dimensional materials, particularly where experimental information is not available.

1.4.2 *Two Dimensional Catalysts for Fuel Cells*

There are numerous examples in the literature of the use of DFT to identify technological interesting materials. For example DFT has been used to identify possible catalysts that can replace platinum in fuel cells. There is a need to develop inexpensive catalysts for the reaction at the cathode. Presently platinum is employed to catalyze the reactions that produce H_2O at the cathode of fuel cells. Platinum is expensive and susceptible to time dependent drift and CO poisoning. These issues are significant obstacles to the development of large scale commercial application of fuel cells. In order to produce H_2O at the cathode atomic oxygen is needed to react with the H^+ ion that is moving from the anode to the cathode. Two possible reactions can provide atomic oxygen.

$$2O + 4H^+ + 4e \rightarrow 2H_2O. \tag{1.28}$$

Another possibility is the formation of O_2H which bonds to the catalyst followed by the removal of OH which could then undergo the following reaction,

$$OH + H^+ + e \rightarrow H_2O. \tag{1.29}$$

There has been some previous work that has indicated that two dimensional molecules such as graphene, carbon nitride and boron

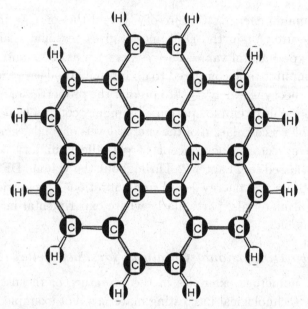

Figure 1.5 Calculated minimum energy structure of nitrogen doped coronene. (Ref. 6)

carbide nitride analogs of graphene could catalyze the reactions at the cathode.[4,5] Nitrogen doped coronene, (CorN), ($C_{23}NH_{12}$) is a planar molecule whose structure is illustrated in Figure 1.5. DFT has been used to show that it has the potential to be a catalyst for the dissociation of O_2H producing OH. The minimum energy structures of $CorNHO_2$ and CorNO are calculated using DFT at the B3LYP/6-31 G* level.[5] The structures are illustrated in Figure 1.6. The bond dissociation energy (BDE) to remove OH from $CorNHO_2$ is then calculated. The BDE is given by,

$$[E((CorNO) + E(HO)] - [E((CorNHO_2)] \qquad (1.30)$$

E is the total electronic energy plus the zero point energy (ZPE) of the minimum energy structure. The ZPE is the total ZPE of all of the normal modes of vibration given by,

$$E_{zpe} = (1/2)h \sum_{i}^{3N-6} f_i \qquad (1.31)$$

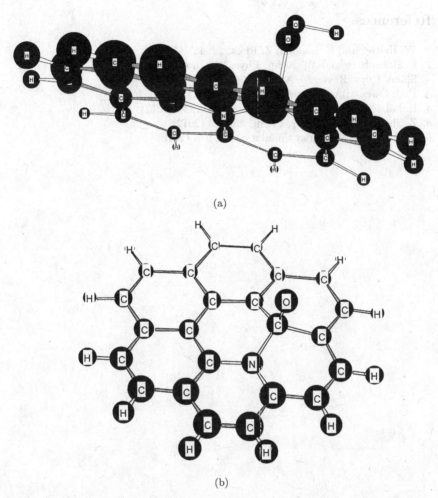

(a)

(b)

Figure 1.6   Calculated minimum energy structure of HO_2 (a) and O (b) bonded to nitrogen doped coronene. (Ref. 6)

where f_I are the vibrational frequencies of the normal modes and N the number of atoms in the molecule. The calculation showed that the BDE to remove an OH from CorNHO$_2$ is 1.47 eV, significantly lower than the calculated BDE to dissociate an HO_2 molecule into O and OH which is calculated to be 5.82 eV.[6] This suggests that $C_{23}NH_{12}$ could be a catalyst for the O_2H dissociation at the cathode of a fuel cell.

References

1. W. Heitler and F. London, Z. Phys. *44*, 455 (1927)
2. P. Hohenberg and W. Kohn, Phys. Rev. *136*, B864 (1964), W. Kohn and L. Sham, Phys. Rev. *140*, A1133 (1965)
3. F. J. Owens (unpublished)
4. L. Lai *et al.* Energy and Environ. Sci. *5*, 7936 (2012)
5. Z. Sheng *et al.* J. Mater. Chem. *22*, 390 (2012)
6. F. J. Owens, Molecular Simulation *42*, 976 (2016)

Chapter 2

Electronic Properties

2.1 Energy Bands of Solids

When atoms, ions and molecules are formed into lattices, whether one two or three dimensional lattices, the discrete energy levels of the substituents form bands of energy. Thus the distinguishing characteristic of the electronic structure of solids is the existence of bands of electronic energy rather than discrete energy levels that exist in atoms or molecules. The top filled energy level of sodium is a 3s state. The first unoccupied energy level is a 3p state. Figure 2.1 illustrates how these states are broadened into bands as the separation of the sodium atoms is reduced to form the body centered cubic lattice of sodium. In the lattice the 3s and 3p levels merge into one broad band. Because the 3p level contains no electrons, this top energy band is not filled and thus sodium is a metal. The energy of the outer electrons can easily be increased by the application of an electric field, and thus metals can conduct electricity.

In the case of insulators there is a large energy gap between the top filled band and the first empty band above it. In the insulating alkali halides, such as NaCl, the gap is 8.6 eV. Semiconductors have much smaller gaps. For example, the semiconductor, germanium has a band gap of only 0.67eV. In this case electrons can be thermally excited to the conduction band.

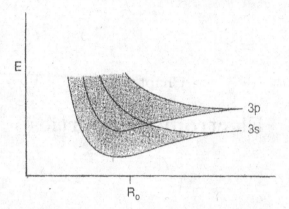

Figure 2.1   Schematic of top half filled 3s energy level and first empty 3p level as a function of sodium atom separation in sodium metal showing the narrowing of the bands as the separation increases.

2.2 The Free Electron Fermi Gas in Low Dimensional Metals

One of the simplest models of the electronic structure of metals treats the conduction electrons as though they see no potential at all, but are confined to the volume of the solid. The model is best applicable to monovalent metals such as lithium, sodium or potassium, where the ion cores only occupy about 15% of the volume of the solid. The energies are obtained by solving the Schrödinger wave equation for $V[r] = 0$, with boundary conditions. For the case of a one dimensional chain of metal atoms, the wave equation has the form,

$$-[h^2/2m]d^2\Psi_n/dx^2 = E_n\Psi_n \qquad (2.1)$$

where Ψ_n is the wave function of the electron in the nth state, E_n is the energy and m is the mass of the electron. The boundary conditions for a one dimensional lattice of length L are,

$$\Psi_n(0) = 0 \quad \text{and} \quad \Psi_n(L) = 0 \qquad (2.2)$$

The eigen values obtained by solving equation (2.1) are,

$$E_n = [h^2/2m][n/2L]^2 \qquad (2.3)$$

where n is a quantum number having integer values $0, 1, 2, 3, \ldots$,etc.

Equation (2.3) is useful in understanding how the electronic structure of a one dimensional metal is affected when the lengths are nanometers. The separation between the energy levels of state n and n+1 is,

$$E_{n+1} - E_n = [h^2/8mL^2][1 + 2n] \qquad (2.4)$$

It is seen from equation (2.4) that as the length of the chain, L, decreases the separation between energy levels increases, and eventually the band structure opens up into a set of discrete levels. This also means that the density of states, the number of energy levels per interval of energy, will decrease with size. When the energy levels are filled with electrons only two electrons are allowed in each level because of the Pauli exclusion principle. These two electrons must have different spin quantum numbers m_s of $+1/2$ and $-1/2$ meaning that the two electron spins in each level n are antiparallel, and there is no net spin in the level. The Fermi energy is the energy of the top filled level which for a monovalent metal will have the quantum number $n_f = N/2$ where N is the number of atoms in the solid. Thus for the one dimensional solid, the Fermi energy is obtained from equation (2.3) to be,

$$E_f = [h^2/2m][N/4L]^2 \qquad (2.5)$$

The energy levels for a two dimensional Fermi gas are given by,

$$E_n = (h^2/2ma^2)(n_x^2 + n_y^2) = E_0 n^2 \qquad (2.6)$$

This can also be expressed in terms of the wave vector $K = 2\pi/\lambda$ where λ is the wave length of the electrons.

$$E_k = (h/2\pi)^2 (K_x^2 + K_y^2)/2m \qquad (2.7)$$

Where $K_x = 2\pi p/L$ and $K_y = 2\pi q/L$ where p and q have the values 1,2,3 etc. and L is the length of the edges of the square unit cell.

2.3 Periodic Boundary Conditions

The use of periodic boundary conditions is possible because of Bloch's Theorem.

The theorm states that for any wave function that satisfies the Schrodinger equation there exists a wave vector, K, such that a translation by a lattice vector a is equivalent to multiplying the wavefunction by a phase factor $\exp(iK \cdot a)$. i.e

$$\Psi_K(r + a) = \exp(iK \cdot a)\Psi_K(r) \tag{2.8}$$

Consider a one dimensional lattice having a lattice vector a. To correctly obtain the energy levels we would have to consider a lattice of length Na where N is infinite. This off course is not computationally possible. Now if N were finite, it would be necessary that the wave function be zero at the ends of the lattice. Because of reflections from the ends, this would lead to standing waves which would have to be included in the solution and do not exist in large crystals.

Periodic boundary conditions or Born–von Karman boundary conditions provide a mathematical device to get around the physical affects of boundaries. In one dimension the device forms the lattice into a circle of N cells. To insure that there is no discontinuity of the wave function it is required that,

$$\Psi(X + Na) = \Psi(X) \tag{2.9}$$

where, a, is the lattice parameter and N is the number of cells in the lattice. The Bloch condition in one dimension is

$$\Psi_K(X + Na) = \exp(iKNa)\Psi_K(X) \tag{2.10}$$

Thus $\exp(iKNa) = 1$ which means $K = 2\pi p/Na$ where p is an integer. In the reduced zone in one dimension K must have values as $-\pi/a < K > \pi/a$. The integers p go from $-(1/2)N$ to $(1/2)N$. The number of allowed wave vectors in the Brillouin zone equals the number of unit cells in the crystal.

2.4 Density of States

An important property of solids is the density of states, that is the number of energy levels per interval of energy given by $D(E) = dN/dE$. The density of states determines a number of properties of solids such as the electronic specific heat and the magnetic susceptibility arising from the conduction electrons. The density of states depends on the dimensionality of the material. For the Fermi gas model of solids, the density of states in one dimension can be obtained from equation (2.5) by solving the equation for N and then taking the derivative with respect to E. This yields,

$$dN/dE = (4L/h)(m/2E)^{1/2}. \tag{2.11}$$

The density of states in two dimensions can be shown to be,

$$dN/dE = L^2\pi m/h^2 \tag{2.12}$$

Notice that the density of states is independent of the energy in two dimensions.

The density of states in 3 dimensions is given by,

$$dN/dE = (V/2\pi^2)(2m\pi^2/h^2)^{3/2}E^{1/2} \tag{2.13}$$

where V is the volume of the cube confining the electrons. It is seen that the dependence of the density of states on E depends on the dimension of the solid. Figure 2.2 gives a plot of the density states for solids of different dimensions.

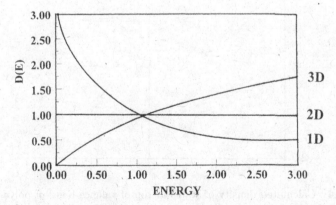

Figure 2.2   Density of electronic states, D(E)=dN/dE in materials of different dimensions.

2.5 Van Hove Singularities

Another characteristic of low dimensional materials is the emergence of singularities in the density of states near the top of the valence band, referred to as Van Hove Singularities. As an example consider the case of the two dimensional polymer chain, polyacetylene whose structure is illustrated in Figure 2.3. Figure 2.4 shows the results of a calculation of the density of states at the top of the valence band

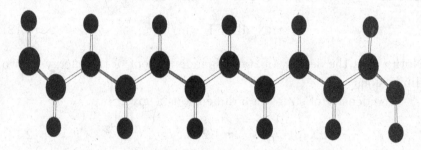

Figure 2.3 Structure of the polyacetylene polymer.

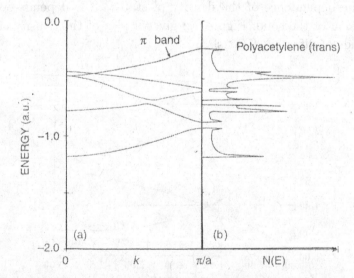

Figure 2.4 Calculated density of states at top of valence band of polyacetylene showing spikes in the density of states characteristic of low dimensional materials. (Ref. 1)

of polyacetylene showing spikes in the density of states.[1] The existence of these spikes can have an important affect on the vibrational frequencies which will be discussed in the next chapter.

2.6 The Tight Binding Model

The tight binding model takes a different approach to treating the electronic structure of metals. It assumes that when the electron is close to the atom of the solid, it will have a wavefunction $\Phi_a(r - l)$ corresponding to the wave function of the free atom. When the electron is far from the atom the wavefunction can be described as that of a free electron, i.e $\exp(iK \cdot l)$. The wave function is written in this model as,

$$\Psi_k[r] = \Sigma_l \exp(iK \cdot l)\Phi_a(r - l) \qquad (2.14)$$

The function looks like a series of strongly localized atomic orbitals whose amplitude is modulated by a phase factor. The model is appropriate to materials having atoms with d orbitals such as transition metals which are compact and form narrow well defined bands. The energy levels are determined by,

$$E = Eo + < \Psi_k^*[r]H\Psi_k[r] > \qquad (2.15)$$

which is,

$$E = Eo + \Sigma_l \exp(iK \cdot l) < \Phi_a(r - l)H\Phi_a(r) > \qquad (2.16)$$

where H is the difference between the potential of the free atom and the potential the atom sees in the crystal, $H = V[r] - V_a[r]$. For the case of a monatomic two dimensional square lattice having lattice parameter, a, and assuming only nearest neighbor interactions, equation (2.16) becomes,

$$E = Eo + 2t(\cos K_x a + \cos K_y a) \qquad (2.17)$$

where t is given by,

$$< \Phi_a(r - l)H\Phi_a(r) > \qquad (2.18)$$

It turns out that t is negative. The $\Phi_a(r)$ for the top occupied band is the highest occupied atomic orbital of the free atom, and E_0 is the energy of the top occupied orbital of the free atom . The lowest energy of the band is $E_0 + 4t$ occurring at $k = 0$ and the highest energy occurs at $E_0 - 4t$ which means the band width is 8t. The separation between the highest and lowest energy depends on t which involves integrals of the atomic orbitals over the potential in the crystal given by equation (2.18). Consider a one dimensional finite lattice of length L having N atoms separated from each other by, a. The number of states in an energy band will be N corresponding to the allowed wavelengths of the wavefunction that fit into L such that $\psi_N(0) = 0$ and $\psi_N(L) = L$. In the tight binding model the dependence of E on K in a band will be,

$$E = Eo - 2tcosKa \qquad (2.19)$$

where K can have values from 0 to $N\pi/L$.

2.7 Electronic Structure of Covalently Bonded Materials

Some of the most technologically interesting materials are solids in which the atoms are covalently bonded to each other. Silicon the work horse of the electronic industry is an important example. Silicon, the major component of most transistor devices, has four valence electrons, which are shared in the covalent bonds with four nearest neighbor silicon atoms in the silicon lattice. Figure 2.5 is an illustration of the unit cell of the silicon lattice showing the four nearest neighbors. Covalent bonding involves the overlap of the outer wave functions of the nearest neighbor atoms as discussed in Chapter 1 on molecular orbital theory. If the lattice is doped with a phosphorous atom, which has five valence electrons, an extra electron is available to contribute to the conductivity of the lattice. The donor energy level of this extra electron lies in the band gap just below the conduction band, as shown in Figure 2.6, such that a very small amount of thermal energy, kT, will excite it to the conduction band. This kind of doped semiconductor is called an N type semiconductor. If

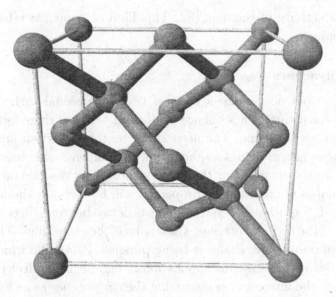

Figure 2.5   Unit cell of the silicon lattice.

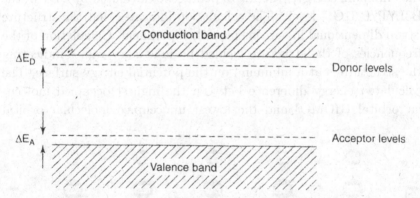

Figure 2.6   Illustration of the effect of N and P doping on energy levels in the band gap of semiconductors.

silicon atoms are replaced by aluminum, which has three valence electrons, one bond will be missing an electron and is referred to as a hole. The acceptor energy level of the hole, as shown in Figure 2.6, is just above the valence band, and this hole can also contribute to the conductivity of the material. This kind of doping is referred to as P doping.

2.8 Polyacenes

Acenes or polyacenes are a class of two dimensional carbon compounds consisting of a sequence of bonded benzene rings forming a two dimensional plane. The acenes can be considered an analog of a graphene nanoribbon. Graphene is a two dimensional structure of carbon atoms arranged in the same structure as the carbon atoms in the planes of graphite. Its properties will be discussed in detail in Chapter 4. Naphthalene, $C_{10}H_8$, consists of two benzene rings and the longest of the series heptacene, $C_{30}H_{18}$ has 7 benzene rings. The possibility of even longer chains is being pursued. Pentacene which consists of 5 benzene rings has application in organic field-effect transistors. Here the interest is in discussing their properties as an example of a covalently bonded two dimensional material. Figure 2.7 shows the minimum energy structure of pentacene calculated by DFT at the B3LYP/6-31G * level.[2] The calculation indicates that the structure is two dimensional in agreement with experiment. Calculation of the frequencies of the structure yielded no imaginary values indicating the structure is at a minimum on the potential energy surface. The calculated energy difference between the highest occupied molecular orbital (HOMO) and the lowest unoccupied molecular orbital

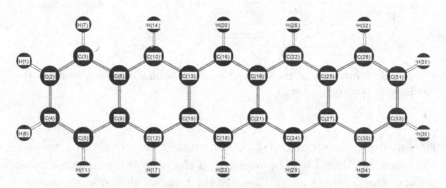

Figure 2.7   Structure of the pentacene molecule.

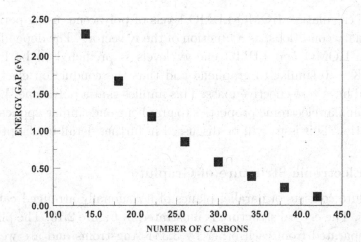

Figure 2.8   Energy gap at the center of the Brillouin zone versus the length of an acene chain. (Ref. 2)

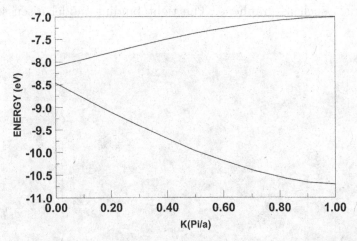

Figure 2.9   Highest occupied energy level and lowest unoccupied level versus the wave vector of acenes. (Ref. 2)

(LUMO), known as the energy gap, is plotted in Figure 2.8 versus the number of carbon atoms in polyacene chains showing that the energy gap approaches zero as the length of the chain increases.[2] This is a general result that has been noted in a number of narrow two dimensional materials such as polyacetylene. Figure 2.9 is a plot of the calculation HOMO-LUMO levels of polyacene, using periodic boundary conditions, as a function of the K vector.[2] The dependence of the HOMO and LUMO energy levels is predicted to be linear near $K = 0$ similar to graphene and thus the conduction electrons should have zero effective mass. This implies that a polyacene should have similar electronic properties to graphene and similar application potential. This issue will be discussed in further detail in Chapter 4.

2.9 Electronic Structure of Graphite

Graphite consists of parallel planes of hexagonally arranged carbon atoms. The crystal structure is illustrated in Figure 2.10. The planes are separated from each other by 3.354 Angstroms and are weakly bonded to each other by a Van der Waals potential. Graphite has played an important role in the development of two dimensional materials such as graphene. The tight binding model discussed in

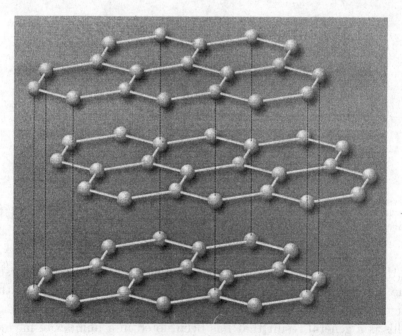

Figure 2.10   Illustration of the structure of graphite.

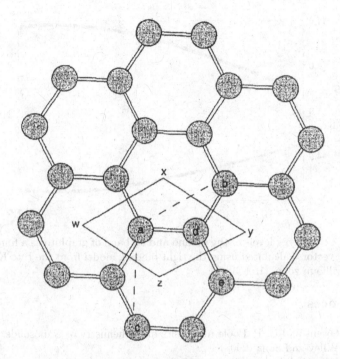

Figure 2.11   Unit cell of graphite.

Section 2.6 can be used to obtain the dispersion relationship assuming only nearest neighbor interactions. The unit cell of graphite is shown in Figure 2.11. The cell is hexagonal delimited by the lines connecting the points w, x, y, z in the figure. The fundamental lattice displacements are $\mathbf{a}_1$ which is the dashed line from a to b and has a length $1.42(3)^{1/2}$ Å $= 2.46$ Å, and $\mathbf{a}_2$ the dashed line from a to c, which has the same length as $\mathbf{a}_1$. The reciprocal lattice vectors have the magnitude $2/a(3)^{1/2}$ and have the directions ab and ae. Thus the reciprocal lattice vectors are in different directions than unit cell directions in real space which complicates obtaining the dispersion relationship. Figure 2.12 is a plot of the dependence of the energy levels versus the wave vector of the σ band and σ^* band calculated by the tight binding model discussed in Section 2.6 from the Γ to K points in the Brillouin zone.[3]

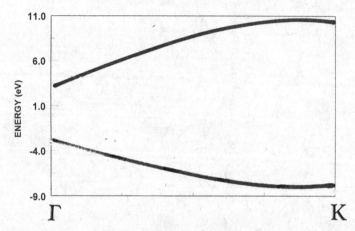

Figure 2.12 Energy levels of the σ band and σ^* band of graphite as a function of the wave vectors calculated using the tight binding model from the Γ to K points in the Brillioun zone. (Ref. 3)

References

1. F. J. Owens and C. P. Poole in Physics and Chemistry of Nanosolids, p. 211, John Wiley and Sons (2008)
2. F.J. Owens Solid State Comm. *185*, 58 (2014)
3. F. J. Owens (unpublished)

Chapter 3

Vibrational Properties

3.1 Measurements of Vibrational Frequencies

The atoms of a molecule or solid lattice vibrate. The specific frequencies, called the normal modes of vibration, are determined by the nature of the interaction between atoms of the molecule or lattice and the symmetry of the entity. The vibrational frequencies of solids can be measured by infra-red (IR) spectroscopy and Raman spectroscopy. IR spectroscopy measures the absoption of IR light when it induces a transition from the $N = 0$ vibrational state to the $N = 1$ state. The basis of Raman spectroscopy is illustrated in Figure 3.1. Laser light is used to excite the lowest energy level of a vibration to some higher level. The higher level excited state then decays back to the lowest level. However, some of decay goes to a vibrational state above the ground state. The frequency of this emitted light is measured and the difference between the frequency of exciting laser light and the emitted light measures the frequency of the vibration.

As an example let us consider measurements of the vibrational frequencies of graphite. Graphite consists of parallel planes of hexagonally arranged carbon atoms. The crystal structure is illustrated in Figure 2.10. The planes are separated from each other by 3.354 Angstroms and are weakly bonded to each other by the Van der Waals potential. Because of the weak interaction between the planes it was possible to synthesize graphene by using scotch tape to remove

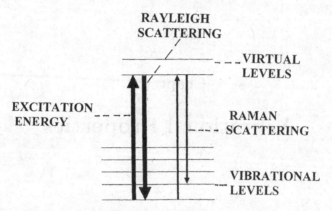

Figure 3.1   Illustration of excitation and emission of light from vibrational energy levels providing the basis of Raman spectroscopy.

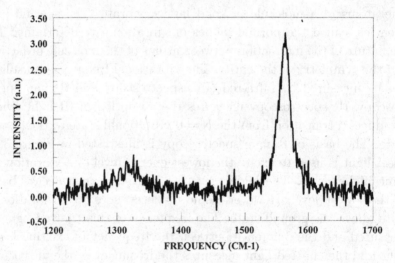

Figure 3.2   Raman spectra of graphite showing the D and G modes (Ref. 1)

a small number of carbon layers from single crystals of graphite. This will be discussed in more detail in Chapter 6.

Figure 3.2 shows the Raman spectra of graphite between 1200 cm^{-1} and 1700 cm^{-1}.[1] Two Raman lines are observed at 1563 cm^{-1} and 1320 cm^{-1}. The Raman line at 1563 cm^{-1} is the E_{2g}, mode commonly referred to as the G mode. It involves a stretch of the

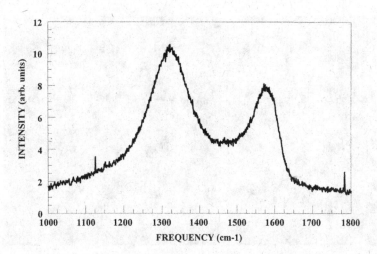

Figure 3.3   Raman spectra of graphite oxide. (Ref. 1)

intra-layer C-C bond lengths. The line at 1320 cm^{-1}, referred to
as the D mode, involves a breathing of the sp^2 C-C bonds in the
carbon rings. The ratio of the intensity of this mode to the G mode,
I_D/I_G, is a measure of the number of defects in the carbon planes.
For the case of the graphite Raman spectra shown in Figure 3.2, this
ratio is 0.17 indicating that the layers in this graphite have very few
defects. There is also a Raman line observed at 2663 cm^{-1} which
is an overtone mode of the D mode referred to as the 2D mode.
Figure 3.3 shows the Raman spectra of oxidized graphite.[1] It is seen
that there is a large increase in the intensity of the D mode relative
to the G mode. This indicates that the chemical process of oxidation
introduces a large number of defects into the graphite planes.

3.2 Surface Enhanced Raman Spectroscopy (SERS)

The Raman effect is quite weak having a cross section of the order
of 10^{-30} cm^2 to 10^{-28} cm^2. The larger cross section occurs when the
energy of the laser is equal to a separation of an electronic energy
level in the molecule. This is referred to as resonant Raman spectroscopy. Thus Raman spectroscopy is not a very sensitive technique.
However, it was found that when Raman spectra were recorded from

Figure 3.4 Picture of gold coated silicon substrate used to obtain SERS spectra taken with 50X lens of the optical microscope of a confocal micro-Raman spectrometer. (Ref. 3)

a layer of molecules deposited on nanostructured metal surface such as a thin film of closely spaced silver nanoparticles, the cross section could be enhanced up to 10^{-16} cm^2 representing a 14 order of magnitude enhancement. It was shown that a film of 8 nm silver particles resulted in a larger enhancement of the vibrations of pyridine compared to a film of 30 nm particles indicating that particle size is an important factor in determining the enhancement.[2] Since the initial discovery, a number of different nanostructured surfaces have been fabricated and shown to produce enhancements of Raman line intensities. Figure 3.4 shows a microscope image of one such structure. The SERS substrate is comprised of a square lattice of inverted square pyramidal pits. The pits were produced by conventional photolithography in a silicon dioxide mask on a (100) oriented silicon wafer.[3] Anisotropic etching was then performed using KOH to preferentially etch the (111) planes to give an array of the inverted square pyramidal pits. A uniform 300 nm layer of gold was sputtered on the wafer to give a smooth coating on the pits and between the pits as well as off the structured region. The coating was smooth off the structured region having no pits allowing comparison of the

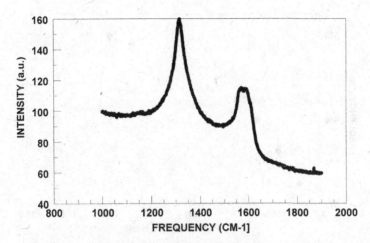

Figure 3.5   SERS spectrum of graphite showing large increase of D mode. (Ref. 1)

Raman spectra of the material on and off the nanostructured region. It has been shown that these substrates provide reproducible surface enhanced Raman spectra of many materials.[3]

Figure 3.5 shows the SERS spectrum of graphite between 1000 cm^{-1} and 1900 cm^{-1} obtained when the graphite is on the structured region shown in Figure 3.4.[1] The enhanced Raman measurements, were made by depositing suspensions of the material in acetone on the substrate. No spectrum was obtained from the drop on the flat gold non structured region. The G mode in the SERS spectrum is observed at 1585 cm^{-1} which represents an upward shift from the bulk graphite at 1563 cm^{-1}.

The SERS effect is a result of the laser used in Raman spectroscopy exciting vibrations of the electron plasma of the metal. In a metal the electrons are not localized on the atoms of the metal and are free to move with the application of an electric field. These free electrons can be viewed as a cloud of electrons which can vibrate as a whole entity. This is referred to as a plasma oscillation. The excited vibrating cloud of electrons produces an intense oscillatory electromagnetic field in the cavities of the gold nanostructured material shown in Figure 3.4. This time varying electric field causes further

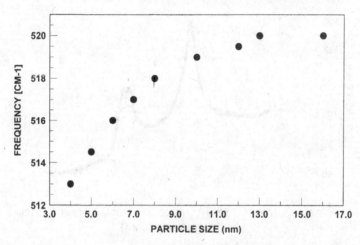

Figure 3.6 Plot of the frequency of the longitudinal optical mode of silicon versus the particle size measured by Raman spectroscopy. (Ref. 4)

excitations from the ground vibrational state of the molecule to the virtual state further enhancing the intensity of the Raman spectra.

3.3 Effect of Size on Vibrational Frequencies of Solids

When solids are reduced to nanometer dimensions the frequencies generally change. Figure 3.6 shows a plot of the decrease in the frequency of the longitudinal optical mode of silicon as a function of particle size measured by Raman spectroscopy.[4] In some instances the frequency increases with reduced particle size. For example the frequency of the acoustic mode of nanoparticles of $CdS_{0.65}Se_{0.35}$ increases as the particle size is reduced.[5] This also occurs in low dimensional materials. Figure 3.7 is a plot of the calculated frequency of acenes as a function of the length of the chain.[6] The acenes were described in Chapter 2. The reason for the change in frequency is phonon confinement. This occurs when the dimensions of the solid are in the order of the wavelength of the lattice vibrations. The uncertainty principle can be used to explain phonon confinement.

The uncertainty principle says that the order of magnitude of the uncertainty in position ΔX times the order of magnitude of the uncertainty in momentum ΔP must at least be, h, Planck's constant

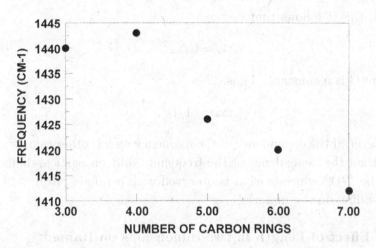

Figure 3.7   Calculated frequency of acene chains versus length (Ref. 6)

divided by 2π, that is,

$$\Delta X \Delta P \geq h/2\pi \qquad (3.1)$$

Let us assume that ΔX is the diameter of the nanoparticle, D, and that it can be measured accurately by some technique such as scanning electron microscopy. This means the uncertainty in the momentum P will have a range of values. It can be shown that the momentum of a phonon is $hk/2\pi$ where k is the wave vector given by ω/c, and c is the velocity of light. Thus we have $D\Delta k \geq 1$ or the minimum uncertainty in k is,

$$\Delta k \sim 1/D \qquad (3.2)$$

Raman spectroscopy measures frequencies at the center of the Brillouin zone, $k = 0$. However, that is a precise value which equation (3.2) shows has some uncertainy, meaning that there is a spread of values for k for a Raman measurement in small particles. This means non-zero values of k will contribute to the Raman spectrum. Let us assume that for small values of k the dependence of the frequency on k, i.e., the dispersion relationship, is quadratic,

$$\Delta \omega = ak^2. \qquad (3.3)$$

From this it follows that,

$$\Delta k = C \Delta \omega^{1/2} \tag{3.4}$$

where C is a constant. Thus,

$$\Delta \omega \sim 1/D^2. \tag{3.5}$$

However, if the dependence of the frequency on k is other than k^2, say k^α, then the dependence of the frequency shift on particle diameter will be $1/D^\alpha$ which is what is observed with α ranging from 1 to 1.5 depending on the material.

3.4 Effect of Length in two dimensions on Raman Intensity

Figure 3.7 shows the structure of a two dimensional polyacene chain. Figure 3.8 is a plot of the calculated intensity of the Raman breathing mode in polyacenes versus the number of carbon rings showing a large increase in intensity as the chain becomes longer.[6] This increase in the intensity of the Raman lines with length is another characteristic of low dimensional materials. It is a result of the emergence of singularities in the density of electronic states near the band gap, called Van Hove singularities, as discussed in Chapter 2. In the Raman experiment the material is electronically excited typically using a laser having light in the visible region of the spectrum. If the excitation occurs from an electronic level where the density of states is high, more electrons are involved in the transition and the Raman spectrum becomes more intense.

3.5 Vibrational Density of States in Low Dimensional Materials

The vibrational density of states $D(\omega)$ is the number of vibrational modes per interval of frequency, $dN/d\omega$. For a one-dimensional line having N atoms the number of vibrational modes is N. For a 3 dimensional lattice it is 3N. The normal modes may be considered a set of independent oscillators with each oscillator having the energy ε_n

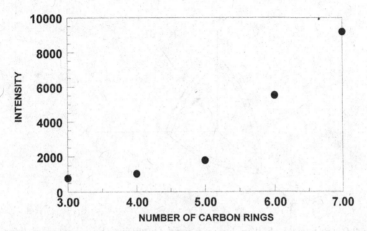

Figure 3.8   Plot of the calculated intensity of the Raman breathing mode of polyacenes versus length showing a marked increase in intensity with increasing length. (Ref. 6)

given by,

$$\varepsilon_n = \left(n + \frac{1}{2}\right)(h/2\pi)\omega \tag{3.6}$$

where n is a quantum number having integral values ranging from $0, 1, 2 \ldots$, to n. The average energy of a harmonic oscillator assuming a Maxwell-Boltzman distribution function is,

$$<\varepsilon> = (h/2\pi)\omega[1/2 + 1/(\exp\{(h/2\pi)\omega/k_BT\} - 1)] \tag{3.7}$$

The total energy of a collection of oscillators is,

$$E = \Sigma_k <\varepsilon_k> (h/2\pi)\omega_k. \tag{3.8}$$

When there is a large number of atoms the allowed frequencies are very close and can be treated as a continuous distribution allowing replacement of the summation by an integral, where $D(\omega)d\omega$ is the number of modes of vibration in the range ω to $\omega + d\omega$. It turns out it is more convenient to work in k space, and deal with the number of modes $D(k)dk$ in the interval k to k + dk. As shown in Figure 3.9 the number of values of k in three dimensions between k and k + dk

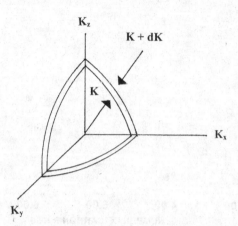

Figure 3.9   Spherical shell in k space used to obtain the vibrational density of states for a 3 dimensional solid.

will be proportional to the volume element on a sphere of radius k which is given by,

$$dV = 4\pi k^2 dk = D(k)dk. \qquad (3.9)$$

The dispersion relationship refers to the dependence of the frequency ω on the k vector. One approximation due to Debye assumes that the relationship is linear which is valid for low values of k, i.e.

$$\omega(k) = uk. \qquad (3.10)$$

From equations (3.9) and (3.10)

$$D(\omega)d\omega = B\omega^2 d\omega \qquad (3.11)$$

where B is a constant and thus,

$$D(\omega) = B\omega^2. \qquad (3.12)$$

In three dimensions the total number of modes is 3N, which means

$$\int_0^{\omega_d} D(\omega)d\omega = 3N \qquad (3.13)$$

where ω_d is the highest frequency that can propagate in the lattice, and is referred to as the Debye frequency. Thus for 3 dimensions in the Debye approximation the density of states for $\omega < \omega_d$ is,

$$D(\omega) = 9N\omega^2/\omega_d^3. \tag{3.14}$$

In two dimensions we would carry out the analogous derivation using a circle having area πK^2 obtaining,

$$D(k)dk = 2\pi kdk \tag{3.15}$$

and

$$D(\omega)d\omega = \omega d\omega \tag{3.16}$$

$$\int_0^{\omega_d} D(\omega)d\omega = 2N \tag{3.17}$$

$$D(\omega) = 4\omega N/\omega_d^2. \tag{3.18}$$

Following the same procedure the density of states in one dimension can be obtained to be,

$$D(\omega) = N/\omega_d. \tag{3.19}$$

Thus the density of phonon states in the Debye approximation depends on the dimensionality of the material. Figure 3.10 shows a plot of phonon density of states in the Debye approximation versus the frequency for materials of different dimensions.

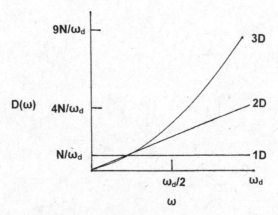

Figure 3.10 Density of phonon states versus frequency in the Debye approximation for 1, 2 and 3 dimensional materials.

References

1. F. J. Owens, Mol. Phys. *113*, 1280 (2015)
2. H. Seki, J. Chem. Phys. *76*, 4418 (1982)
3. M. B. Perney *et al.* Optics Express, *14*, 847 (2006), F.J. Owens, Mol. Phys. *109*, 2587 (2011)
4. Z. Iqbal and S. Vepre, J. Phys. *C15*, 377 (1982), *The Physics and Chemistry of Nanosolids,* p200 by Frank J. Owens and Charles Poole, 2008, John Wiley and Sons, Hoboken, NJ
5. P. Verma *et al.* J. Appl. Phys. *88*, 4109 (2000)
6. F. J. Owens, Solid State Comm. *185*, 58 (2014)

Chapter 4

Carbon Nanotubes

4.1 Single Walled and Multiwalled Carbon Nanatubes

A carbon nanotube can be envisioned as single sheet of graphite rolled into a tube with atoms at the end of the sheet forming the bonds that close the tube. Figure 4.1(a) shows the structure of a single walled carbon armchair nanotube formed by rolling the graphite sheet about an axis in the plane of the sheet perpendicular to a C- C bond. A single walled nanotube (SWNT) can have a diameter of 2 nm and a length of 100 microns, making it effectively a one dimensional structure called a nano-wire. There also exists multi walled carbon nanotubes. An example of a double walled carbon nanotube is shown in Figure 4.1(b).

4.2 Fabrication

Carbon nanotubes can be made by laser evaporation, carbon arc methods, and chemical vapor deposition.

Chemical vapor deposition is the method with the most promise to produce large quantities of carbon nanotubes. Figure 4.2 is an illustration of the equipment needed to make carbon nanotubes by chemical vapor deposition. A quartz glass tube is located in a cylindrical oven. On the bottom of the tube at the center of the oven is a wafer of silicon which has on it a material such as $FeSO_4$. Carbon monoxide, hydrogen and argon are flowed through the glass tube at

41

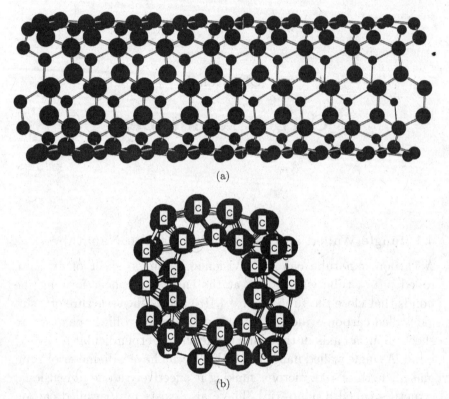

Figure 4.1   (a) Structure of a single walled armchair carbon nanotube and (b) the end view of a double walled carbon nanotube.

very specific flow rates as the temperature of the oven is raised in a controlled manner to $1100°C$. Other gases such as methane can be used as the source of carbon. The hydrogen reduces the $FeSO_4$ to iron nanoparticles. Cobalt and nickel can also be used. The incoming CO gas decomposes into atomic carbon, which assembles on the iron nanoparticles into carbon nanotubes.

Generally when nanotubes are synthesized, the result is a mix of tubes of which two thirds are semiconducting and one third metals. Methods have been developed to separate the semiconducting from the metallic nanotubes. In one approach the separation is accomplished by depositing bundles of nanotubes, some of which are metallic and some semiconducting, on a silicon wafer. Metal electrodes

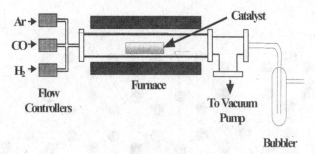

Figure 4.2   Oven used to synthesize single walled carbon nanotubes by chemical vapor deposition.

are then deposited over the bundle. Using the silicon wafer as an electrode a small bias voltage is applied which prevents the semiconducting tubes from conducting, effectively making them insulators. A high voltage is then applied across the metal electrodes, sending a high current through the metallic tubes but not the insulating tubes. This causes the metallic tubes to vaporize, leaving behind only the semiconducting tubes.

4.3 Structure of Carbon Nanotubes

Although carbon nanotubes are not actually made by rolling single graphite sheets into tubes, it is possible to explain the different structures by consideration of the way graphite sheets might be rolled into tubes. A nanotube can be envisioned to be formed by rolling a graphite sheet about a vector labeled T called the translational vector as shown on the top of Figure 4.3. The figure shows examples of two such vectors, T_1 and T_2 having different orientations about which the graphite sheet can be rolled to produce the armchair and the zigzag tubes. Folding about the vector labeled T_2 which is perpendicular to the edges of the hexagons, produces the armchair structure labeled (a) in the figure. Folding about the vector labeled T_1 which is parallel to the sides of the hexagons produces the zig-zag structure labeled (b) in the figure. Folding about any other direction produces chiral tubes. The T_1 vector has the co-ordinates $(n,0)$ and the T_2 vector has the coordinates (n,n). The carbon nanotubes are metallic

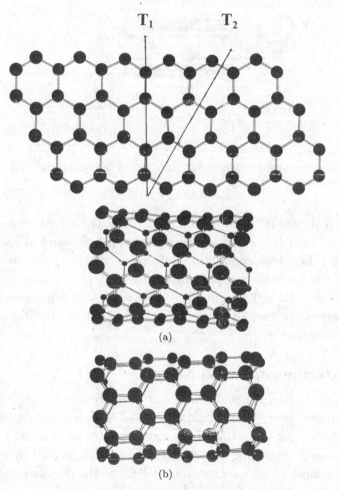

Figure 4.3   A graphite sheet (top) showing vectors T_1 and T_2 about which the sheet can be folded to produce armchair tubes (a) and zig-zag tubes (b).

or semiconducting depending on the diameter and the particular T vector they are rolled about. The metallic tubes have the arm chair structure shown in Figure 4.3a.

4.4 Electronic Properties

Figure 4.4 shows the unit cell of a (5,5) armchair carbon nanotube used to calculate the dependence of the highest occupied

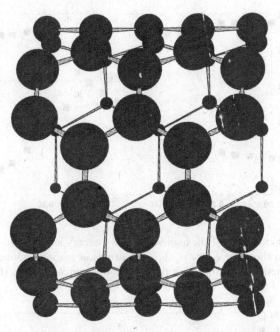

Figure 4.4   Unit cell of (5,5) armchair carbon nanotube used to calculate the dependence of the energy levels on the wave vector.

molecular orbital (HOMO) and the lowest unoccupied molecular orbital (LUMO) on the wave vector K.[1] The K vector is along the axis of the cylindrical tube. Figure 4.5 shows this calculated dependence. Note the crossing of the levels at $K_x \pi/a = 0.65$. Figure 4.6 is a plot of the energy gap of a semiconducting chiral carbon nanotube at the center of the zone versus the diameter, showing that as the diameter of the tube increases the band gap decreases.[2]

Scanning tunneling microscopy (STM), has been used to investigate the electronic structure of carbon nanotubes. A schematic of a scanning tunneling microscope is illustrated in Figure 4.7. A scanning tunneling microscope utilizes a wire with a very fine point. This fine point is positively charged and acts as a probe when it is lowered to a distance of about 1 nm above the surface under study. Electrons at individual surface atoms are attracted to the positive charge of the probe wire and jump (tunnel) up to it, thereby generating a weak

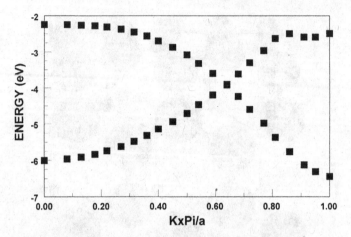

Figure 4.5 Dependence of the highest occupied energy level and the lowest unoccupied level on the wave vector of the (5,5) armchair carbon nanotube calculated by density functional theory using periodic boundary conditions. (Ref. 1)

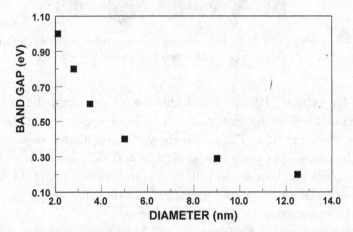

Figure 4.6 The band gap at K = 0 of a semiconducting carbon nanotube versus the tube diameter. (Ref. 2)

electric current. The probe wire is scanned back and forth across the surface in a raster pattern, in either a constant height mode, or a constant current mode. In the constant current mode a feedback loop maintains a constant probe height above the sample surface profile, and the up and down probe variations are recorded. This mode of

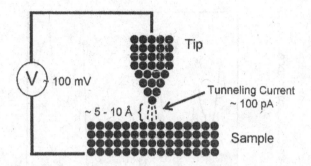

Figure 4.7   A schematic of a scanning tunneling microscope.

operation assumes a constant tunneling barrier across the surface. In the constant probe height mode the tip is constantly changing its distance from the surface, and this is reflected in variations of the recorded tunneling current as the probe scans. The feedback loop establishes the initial probe height, and is then turned off during the scan. The scanning probe provides a mapping of the distribution of the atoms on the surface. To measure the density of states of the carbon nanotube, the position of the STM tip is fixed above the nanotube, and the voltage V between the tip and the sample is swept while the tunneling current I is monitored. The measured conductance $G = I/V$ is a direct measure of the local electronic density of states. Figure 4.8 gives the STM data plotted as the differential conductance, which is $(dI/dV)/(I/V)$, versus the applied voltage between the tip and carbon nanotube for semiconducting tubes.[2] The data show clearly the energy gap in materials at voltages where very little current is observed. The voltage width of this region measures the gap, which for this semiconducting tube is 0.7 eV.

At higher energies sharp peaks are observed in the density of states, above and below the band gap, which are referred to as van Hove singularities, and they are characteristic of low dimensional conducting materials. As we have discussed earlier, electrons in the quantum theory can be viewed as waves. If the electron wavelength is not a multiple of the circumference of the tube it will destructively interfere with itself, and therefore only electron wavelengths, which are integer multiples of the circumference of the tubes are allowed.

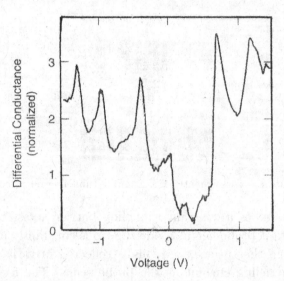

Figure 4.8 Differential conductance versus voltage measured using a scanning tunneling microscope. (Ref. 2)

This severely limits the number of energy states available for conduction around the cylinder. The dominant remaining conduction path is along the axis of the tubes, making carbon nanotubes function as one-dimensional quantum wires. The electronic states of the tubes do not form a single wide electronic energy band, but instead split into one dimensional sub bands which are evident in the data in Fig. 4.8.

In the metallic state the conductivity of the nanotubes is very high. It is estimated that they can carry a billion amperes per square centimeter. Copper wire fails at one million amperes per square centimeter because resistive heating melts the wire. One of the reasons for the high conductivity of the carbon tubes is that they have very few defects to scatter electrons, and thus a very low resistance. High currents do not heat the tubes the way they heat copper wires. Nanotubes also have a very high thermal conductivity, almost a factor of two more than diamond. This means that they are also very good conductors of heat.

Because of their very high electrical conductivity carbon nanotubes have a number of interesting application possibilities. The high electrical conductivity of carbon nanotubes means that they will be poor transmitters of electromagnetic energy. A plastic composite of carbon nanotubes could provide light weight shielding material for electromagnetic radiation. This is a matter of much concern to the military, which is developing a highly digitized battlefield for command, control and communication. The computers and electronic devices that are a part of this system need to be protected from weapons that emit electromagnetic pulses.

The feasibility of designing field effect transistors (FET), the switching components of computers, based on semiconducting carbon nanotubes connecting two gold electrodes, has been demonstrated. An illustration of the device is shown in Figure 4.9. When a small voltage is applied to the gate, the silicon substrate, current flows through the nanotube between the source and the drain. The device is switched on when current is flowing, and off when it is not. It has been found that a small voltage applied to the gate can change the conductivity of the nanotube more than a factor of a million, which is comparable to silicon field effect transistors. It has been estimated that the switching time of these devices will be very fast, allowing clock speeds of a terahertz, which is 10^4 times faster than present processors. The gold sources and drains are deposited by lithographic methods, and

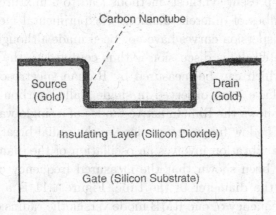

Figure 4.9   Concept of a field effect transistor using carbon nanotubes.

the connecting nanotube wire is less than one nanometer in diameter. This small size should allow more switches to be packed on a chip. It should be emphasized that these devices have been built in the laboratory one at a time, and methods to produce then cheaply in large scale and connected to each other on a chip will have to be developed before they can be used in applications such as computers.

When a small electric field is applied parallel to the axis of a nanotube electrons are emitted at a very high rate from the ends of the tube. This is called field emission. The effect can easily be observed by applying a small voltage between two parallel metal electrodes and spreading a composite paste of nanotubes on one electrode. A sufficient number of tubes will be perpendicular to the electrode so that electron emission can be observed. One application of this effect is the development of flat panel displays. Television and computer monitors use a controlled electron gun to impinge electrons on the phosphors of the screen, which then emit light of the appropriate colors. Flat panel displays using the electron emission of carbon nanotubes are presently under development. A thin film of nanotubes is placed over control electronics with a phosphor coated glass plate on top.

4.5 Vibrational Properties

Because the present synthesis methods lead to a mixture of types of tubes and tubes of different diameters, experimental measurements of phonon dispersion curves have not been made although they have been calculated. This discussion is thus confined to the frequencies at $K = 0$ which can be measured by Raman spectroscopy. No IR frequencies have been observed in single walled carbon nanotubes. Figure 4.10 shows the Raman active modes of a single walled carbon nanotube[3] The low frequency modes are the radial breathing modes (RMB). The vibration involves an oscillation of the diameter of the tube. It has been shown that the measured frequency can be used to estimate the diameter of the tube. Figure 4.11 is a plot of the measured frequency of one RMB mode versus the radius of a (10,10) armchair tube between 3 and 7 angstrom diameters.[2] The Raman

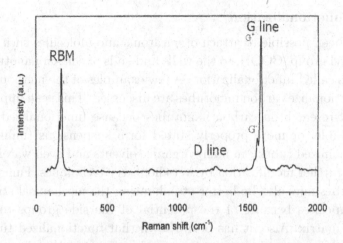

Figure 4.10   Raman spectra of a single walled carbon nanotube. (Ref. 3)

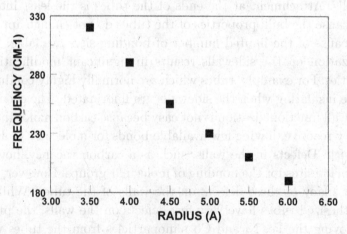

Figure 4.11   Frequency of the radial breathing mode versus carbon nanotube diameter. (Ref. 2)

line labeled the G mode involves a stretch of the C-C bond lengths parallel to the surface of the tube. The lower frequency line, denoted the D mode is a breathing of the sp^2 C-C bonds in the carbon rings. The ratio of the intensity of this mode to the G mode, I_D/I_G, is a measure of the number of defects on the surface of the tubes. The more defects the larger the ratio.

4.6 Functionalization

It has been possible to attach other atoms and molecules such as the carboxyl group, COOH, to the walls and ends of carbon nanotubes, a process called functionalization. A few examples of the methods used to functionalize carbon nanotubes are discussed. This is an important area of research on carbon nanotubes because functionalized tubes are soluble, or more properly stated form suspensions (unlike non functionalized tubes), in many organic solvents and even water. This allows further modification by wet chemistry techniques. Functionalized tubes may also be better at enhancing the mechanical strength of composites because of the potential of the side groups to bond to the matrix. Also, it has been shown that functionalized tubes in some instances tend to aggregate less than non-functionalized tubes. Molecular groups can be attached to the ends of the tubes or the sidewalls. Attachment at the ends of the tubes is the least interesting because the bulk properties of the tubes do not change appreciably because of the limited number of bonding sites available. Functionalization on the sidewalls results in significant modification of properties. For example, tubes which are normally highly conducting become insulating when the sidewalls are fluorinated. The chemistry of sidewall functionalization is not easy because carbon nanotubes are not very reactive, having few available bonds for molecular groups to bind with. Defects in the walls, such as a carbon vacancy, however, can provide sites for the bonding of molecular groups. However, there are not many of these defects on the walls of the tubes. While the as synthesized tubes have very few defects on the walls, the process of removing the Fe, Ni and Co nanoparticles from the tubes which involves treatment with nitric acid can produce defects in the sidewalls such as the rupture of C-C bonds.

A number of different methods have been used to functionalize carbon nanotubes. Nitration and addition of COOH groups has been achieved by suspending carbon nanotubes in concentrated nitric acid while applying sonication for one hour at 70°C. Figure 4.12 shows the Raman spectrum of the tangential modes of the nitric acid treated

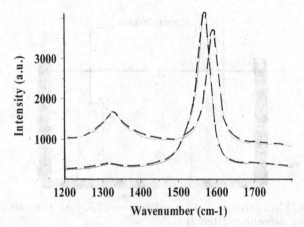

Figure 4.12 Raman spectrum of G mode of single walled carbon nanotubes before (more intense spectrum) and after treatment with nitric acid. (Ref. 3)

carbon nanotubes showing an upward shift of the mode and a reduction in intensity indicative of functionalization.[3] Note also that the intensity of the tangential mode in the functionalized tubes is less than in the non-functionalized tubes. This is the result of a reduction of resonant enhancement because when functionalized, the tubes are less one-dimensional. This is analogous to the reduction of the intensity of the Raman active modes of polyacetylene as the chains become shorter, as discussed in Chapter 2. Also the ratio of the intensity of the D mode to the G mode has increased indicating that the nitric acid treatment increased the number of defects on the surface of the tubes.

The IR spectra shows that a high degree of COOH functionalization has occurred as evidenced by the strong COOH peaks.[3] Evidence of NO_2 vibrations is also present indicating nitration of the carbon nanotubes. The nitric acid treated tubes are soluble in water or form dispersed suspensions, a further evidence that they are functionalized.

Electrochemical methods can also be used to functionalize carbon nanotubes. Nitration of tubes has been achieved by using carbon nanotube paper as the positive electrode in a chemical cell illustrated in Figure 4.13 in which the electrolyte is a 6 molar aqueous

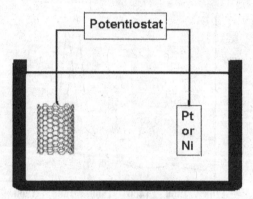

Figure 4.13   Electrochemical cell used to bond NO_2^- to the surface of single walled carbon nanotubes. (Ref. 2)

solution of KNO_2 and the negative electrode is platinum or nickel.[4] Nitration is accomplished by applying one volt across the electrodes for 3 to 4 hours. The Raman spectra of the tangential mode of the nitrated carbon nanotubes shifted down by 3 cm^{-1} compared to the pristine tubes, and the intensity of the spectrum was reduced by a factor of 6 to 7 because of the increase in the dimensionality from one because of the side wall chemical groups.(as discussed above). Figure 4.14 shows the FTIR spectra of the SWNT paper before (top) and after treatment in the electrochemical cell having the KNO_2 solution.[4] The untreated SWNTs show no IR spectrum whereas a number of lines are observed in the treated material. The peak at 1500 cm^{-1} can clearly be assigned to the asymmetric stretch of NO_2 providing evidence of nitration of the tubes.

4.7   Mechanical Properties

Carbon nanotubes are very strong. If a weight W is attached to the end of a thin wire nailed to the roof of a room, the wire will stretch. The stress S on the wire is defined as the load, or the weight per unit cross sectional area A of the wire,

$$S = W/A \tag{4.1}$$

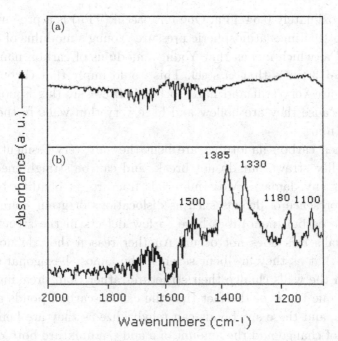

Figure 4.14   FTIR spectrum of carbon nanotube paper before (top) and after treatment in electrochemical cell. (Ref. 4)

The strain e is defined as the amount of stretch, ΔL, of the wire per unit length L_0,

$$e = \Delta L/L_0 \qquad (4.2)$$

where L_0 is the length of the wire before the weight is attached. Hooke's law says that the increase in the length of the wire is proportional to the weight at the end of the wire. More generally we say stress S is proportional to strain e,

$$S = Ee \qquad (4.3)$$

The proportionality constant $E = L_0 W/A\Delta L$ is Young's Modulus, and it is a property of a given material. It characterizes the elastic flexibility of a material. The larger the value of Young's modulus is the less flexible the material. Young's modulus of steel is about 30,000 times that of rubber. Carbon nanotubes have Young's moduli of approximately 0.64 TPa. One terapascal (TPa) is a pressure very close to 10^7 times atmospheric pressure. Young's modulus of steel is 0.21 TPa, which means that Young's modulus of carbon nanotubes is almost 3 times that of steel. This would imply that carbon nanotubes are very stiff and hard to bend. However, this is not quite true because they are hollow and have very thin walls in the order of 0.34 nm.

When carbon nanotubes are bent they are very resilient. They buckle like straws but do not break, and can be straightened back without any damage. Most materials fracture on bending because of the presence of defects such as dislocations or grain boundaries. Because carbon nanotubes have so few defects in the structure of their walls this does not occur. Another reason they do not fracture is that as they are bent severely the almost hexagonal carbon rings in the walls change their structure rather than breaking. This is a unique result of the fact that the carbon-carbon bonds are sp^2 hybrids, and these sp^2 bonds can rehybridize as they are bent. The degree of change and the amount of s and p admixture both depend on how much the bonds are bent.

Strength is not the same as stiffness. Young's modulus is a measure of how stiff or flexible a material is. Tensile strength is a measure of the amount of stress needed to pull a material apart. The tensile strength of carbon nanotubes is about 37 billion Pascals. High strength steel alloys break at about 2 billion Pascals. Thus carbon nanotubes are about 19 times stronger than steel.

Nested nanotubes also have improved mechanical properties, but they are not as strong as their single walled counterparts. For example, multi-walled nanotubes of 200 nm diameter have a tensile strength of 0.007 TPa (that is 7 GPa), and a modulus of 0.6 TPa.

4.8 Carbon Nanotube Polymer Composites

Increasing the strengths of materials such as plastics and metals by incorporation of long carbon fibers such as polyacrylonitrile (PAN) is an established technology. The factors that determine the degree of enhancement of the strength of materials when fibers are incorporated into them are; (1) the ratio of the length of the fiber to the diameter, called the aspect ratio; (2) the ability of the fiber to bond to the material; (3) the alignment of the fiber in the material; (4) the inherent tensile strength of the fiber itself, and (5) a good dispersion of the fibers in the material. Since carbon nanotubes have the highest tensile strength of any known fiber and the largest length to diameter ratio, they should be the ultimate reinforcing fiber and this is one of the important applications of carbon nanotubes. Addition of carbon nanotubes to such polymers as polystyrene, and poly vinyl alcohol has been shown to increase the yield strength, hardness and electrical and thermal conductivity of the polymers. A polyacrylonitrile (PAN)-carbon nanotube composite has been fabricated by suspending PAN and 3 to 20% by weight carbon nanotubes in an organic solvent. The organic liquid is allowed to slowly evaporate while being subjected to sonication. Next the residue is dried for a few hours at 80°C to remove any remaining liquid. The residue is then pressed into a pellet. Both non functionalized and functionalized carbon nanotubes have been incorporated into the polymer by this process. Raman measurements of the tangential mode of the carbon nanotube in the composite show that it has been shifted up from 1571 cm^{-1} in the pristine tubes to 1576 cm^{-1} in the composite indicating interaction between the tube and the polymer. In composites of PAN, made with fluorinated carbon nanotubes, the frequency of the tangential mode shifted down by 10 cm^{-1}. Hardness is measured by placing a weighted indenter on the sample. The harder the material, the less the indenter sinks into the sample. It has been shown that hardness correlates to yield strength. Figure 4.15 is a plot of the hardness versus the percent weight of non-fluorinated and fluorinated carbon nanotubes in PAN showing that the hardness increases with higher percentage of the nanotubes in the composite.[5] The fluorinated tubes are more effective in increasing the hardness. The observed increase of the hardness is not optimal because the tubes have not been preferentially aligned in the polymer. Further even though sonication is applied, the nanotubes remain bundled to some extent. This results in a reduction in the degree of enhancement

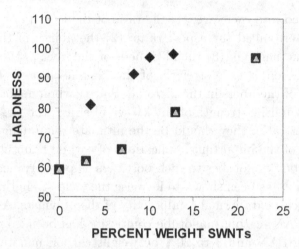

Figure 4.15   Hardness versus percent weight of fluorinated (top curve) and non fluorinated carbon nanotubes in PAN. (Ref. 5)

of hardness because the aspect ratio of the bundles is lower than the isolated tubes and because there may be some slippage of the tubes with respect to each other in the bundles under stress. The enhancement of hardness using the fluorinated tubes may be due to intertube bonding in the bundles and or better bonding to the polymer matrix. Thus while carbon nanotubes have large potential to enhance the mechanical strength of polymers, optimum enhancement has not yet been achieved.

Because the non-functionalized tubes have high thermal and electrical conductivity, incorporating the tubes into polymers can be used to make electrical and thermally conducting polymer materials, which can have a number of applications. Figure 4.16 shows the Raman spectrum of the G mode of single walled carbon nanotubes in the polymer, PVDF, $(CF_2\text{-}CH_2)_N$ showing an increase in frequency relative to the frequency of the pristine tubes in the polymer indicative of bonding to the polymer.[6] Figure 4.17 shows a measurement of the current voltage relationship of PVFD containing 5% by weight single walled carbon nanotubes.[6] The measurement is made with four connections to a disk of the material. Two of the connections are connected to a current source and the other two

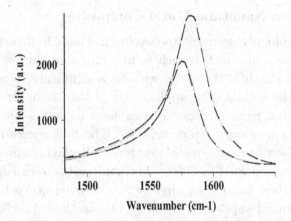

Figure 4.16 Raman spectrum of G mode of single walled carbon nanotubes in poly vinyl diflouride (PVDF) compared to pristine tubes.(larger spectrum). (Ref. 6)

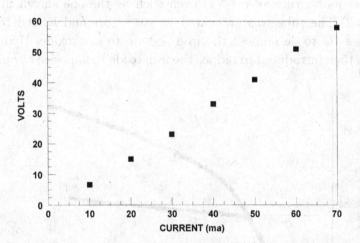

Figure 4.17 Current versus voltage for single walled carbon nanotubes PVDF composites. (Ref. 6)

are to a voltmeter. The four-probe technique eliminates the effect of the contact resistance between the wires and the sample. The voltage versus current is a straight line indicating that the conductivity obeys Ohm's law. The measured resistivty is 3.6 ohm-cm.

4.9 Carbon Nanotube Metal Composites

It would be highly desirable to develop methods to incorporate carbon nanotubes into metals such as iron and aluminum and increase the yield strength of the metal without a significant increase in the weight of the metals. The applications of such stronger metals are many ranging from stronger metals for car bodies and airplanes as well as armor on military vehicles. The first successful fabrication of carbon nanotube-metal composite displaying enhanced yield strength was reported in 2006.[7] The composites were fabricated by growing carbon nanotubes directly into low density pellets of iron by the chemical vapor deposition (CVD) method described earlier. The density of the iron pellets was controlled by the pressure used to press the pellets. The catalysts such as iron acetate were mixed with the iron powder used to make the pellet. The pellets were placed in the quartz tube of a CVD oven such as the one shown in Figure 4.2. The tube was evacuated to 10^{-3} Torr and heated to 500 degrees °C to decompose the iron acetate to iron oxide. Hydrogen gas is then introduced to reduce the iron oxide to nanosized iron. The

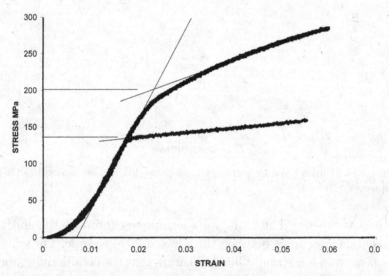

Figure 4.18   Stress-strain curves of carbon nanotubes iron composites (top curve) compared to iron of same density without tubes. (Ref. 7)

temperature is then raised to 700°C and carbon monoxide is flowed through the tube for 30 minutes. The carbon monoxide decomposes into atomic carbon which then forms carbon nanotubes about the iron nanoparticles in the pores of the pellet. Reference pellets without the catalysts which do not produce carbon nanotubes were also subjected to the same treatment. Figure 4.18 shows the stress-strain curve of the carbon nanotube metal composite (upper curve) and the reference pellet of iron.[7] The yield strength is the value of stress where the stress-train curve deviates from linearity indicated by the horizontal lines in Figure 4.18. A 45% increase in yield strength compared to the reference sample is obtained in an iron pellet containing 1% by weight of single walled carbon nanotubes. The SEM picture in Figure 4.19 provides a possible explanation for the enhancement of strength. The figure is a high resolution image of the inside of a pore of the iron composite.[7] It is seen that the carbon nanotubes

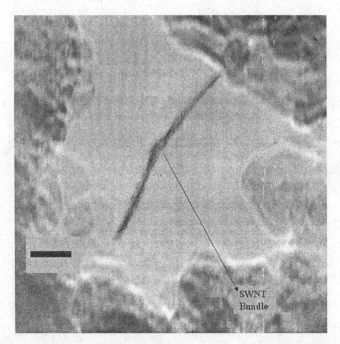

Figure 4.19   SEM image of of iron carbon nanotube composites showing a pore with a carbon nanotube connecting its walls. (Ref. 7)

stretch across the pore and are bonded to the walls of the pore, which contain the iron nanoparticles. It may be that the enhanced strength of the composite is due to the mechanical support provided to the pores by the bonding of the tubes to the walls of the pores.

References

1. M. Miller and F. J. Owens (unpublished)
2. *The Physics and Chemistry of Nanosolids* by Frank J. Owens and Charles Poole, John Wiley & Sons, Hoboken, N.J. (2008)
3. F. J. Owens (unpublished)
4. Y. Wang *et al.* Chem. Phys. Lett. *407*, 68 (2005)
5. F. J. Owens, Mater. Lett. *59*, 3720 (2005)
6. F. J. Owens, J. R. P. Javaakody and S. G. Greenbaum, Composites Sci. and Tech. *66*, 280 (2006)
7. A. Goyal, D. A. Wiegand, F. J. Owens and Z. Iqbal, Mater. Res. *21*, 1 (2006)

Chapter 5

Other Kinds of Nanotubes

5.1 Boron Nitride Nanotubes

Structurally boron nitride nanotubes (BNNTs) resemble carbon nanotubes. Figure 5.1 illustrates the structure of (5,5) armchair BNNT.

5.1.1 *Fabrication*

The BNNTs were first synthesized by introducing boron nitride between two arcing tungsten electrodes.[1] It was not possible to control tube diameter, chirality or the number of nanotubes that nest within each other. Figure 5.2 shows a TEM image of a 4 walled BNNT made by this process.[2] Subsequently many other processes to make the BNNTs were developed such as laser vaporization and chemical vapor deposition. The chemical vapor deposition method is similar to that used to make carbon nanotubes. The apparatus for this method is illustrated in Figure 4.2.

Single walled BNNTs were produced by a laser ablation technique.[3] A 248 nm eximer laser was focused on a substrate on which was deposited a mixture of 98% boron nitride 1% Ni and 1% Co. The substrate was in an oven in which various gases such as argon, nitrogen and helium were used as carrier gases. These gases carried the laser ablated material to a cooled copper collector on which the material condensed. The method produced a large amount of single walled BNNTs.

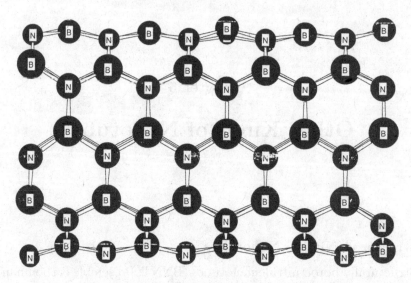

Figure 5.1   Structure of a (5,5) armchair boron nitride nanotube.

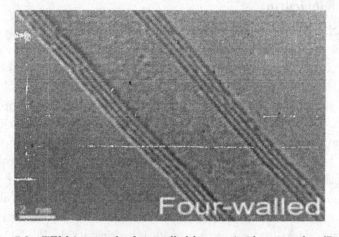

Figure 5.2   TEM image of a four walled boron nitride nanotube. (Ref. 2)

5.1.2 *Electronic Properties*

Figure 5.3 shows the results of a calculation of the top filled energy level (HOMO) and the two lowest unoccupied levels (LUMO) versus the K vector for (4,4) BNNTs.[4] The K vector is along the axis of the

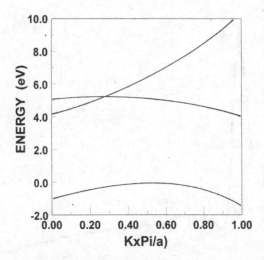

Figure 5.3   Highest occupied energy level and two lowest unoccupied levels versus K vector of a boron nitride nanotube. (Ref. 4)

tube. The calculated band gap at K = 0 is close to 4.6 eV. So the BNNT tubes can be classified as wide band gap semiconductors. The lowest conduction band has a parabolic dependence on the K vector resembling a free electron conductor. Charge carriers introduced into this band by doping or other means are localized inside the tubes along the axis of the tube. Unlike carbon nanotubes the band gap is independent of tube diameter and chirality. Figure 5.4 shows the measured optical absorption spectrum of BNNTs showing a strong absorption at 4.8 eV which corresponds to the band gap of the tubes.[5] Interestingly the theoretically calculated value is in good agreement with the experimental value. It has been found that the band gap can be decreased by application of a dc electric field perpendicular to the axis of the tubes.[6] Figure 5.5 shows the effect of an electric field on the band gap of the BNNTs of two different diameters. The line having the stepper slope is for a tube having a diameter of 49.9Å while the line with the smaller slope is for a tube with diameter of 22.2Å. The larger diameter tubes are more sensitive to the effect of the electric field.

Density functional calculations indicate that increasing the boron content relative to the nitrogen content of zigzag and armchair

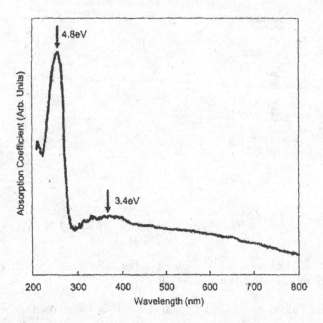

Figure 5.4   Optical absorption spectrum of boron nitride nanotubes. (Ref. 5)

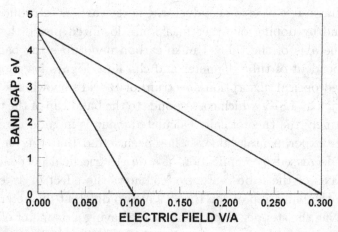

Figure 5.5   Plot of the effect of an electric field on the band gap of boron nitride nanotubes at K=0. The line having the larger slope is for a tube of diameter 49.9Å while the line with the smaller slope is for a tube of 22.2Å. (Ref. 6)

BNNTs can reduce the band gap at K = 0 so that the tubes are small band gap semiconductors.[7] A (5,5) armchair tube, $B_{51}N_{49}$, and a (7,0) zigzag tube, $B_{50}N_{48}$, were predicted to have a band gap of 1.6 eV. These calculations suggest another approach to tuning the band gap of the BNNTs tubes. Theoretical calculations have also shown that introducing defects into BNNTs such as removing one boron atom can significantly reduce the band gap.[8] For example removal of one boron atom in 80 atoms in a (5,5) armchair tube reduces the band gap to 0.95 eV. It was also predicted that zigzag BNNTs having a defect would be ferromagnetic.

BNNTs like carbon nanotubes emit electrons when a dc electric field is applied parallel to the axis of the tubes. Unlike carbon nanotubes the BNNTs display stable emission.[9] Also the emission current in BNNTs follows the Fowler-Nordheim law which predicts a linear relationship between $\ln(I/V^2)$ and $1/V$ where I is the current and V the voltage. Because of these properties BNNTs may be better candidates for use in flat panel displays.

5.1.3 *Vibrational Properties*

The Raman spectrum of the BNNTs is some what different from that of carbon nanotubes. No radial breathing modes are observed and only the E_g mode is observed at $1370\,\text{cm}^{-1}$ considerably lower in frequency that the E_g mode in carbon nanotubes. The E_g mode in the BNNTs is shifted up by $5\,\text{cm}^{-1}$ compared to its frequency in hexagonal boron nitride (H-BN) and is asymmetrically broadened.[10] Unlike carbon nanotubes the BNNTs are IR active. Figure 5.6 shows the IR spectrum.[11] There is a strong absorption at $1382\,\text{cm}^{-1}$ which is due to the B-N stretching vibration and an absorption at $784\,\text{cm}^{-1}$ due to the B-N-B bending vibration. The absorption at $3391\,\text{cm}^{-1}$ is due to water absorbed on the sample.

5.1.4 *Mechanical Properties*

The atomic force microscope (AFM) has been used to measure the mechanical properties of a single SWNT and BNNT. Figure 5.7 is an illustration of the AFM. It consists of a sharp tip having a radius of

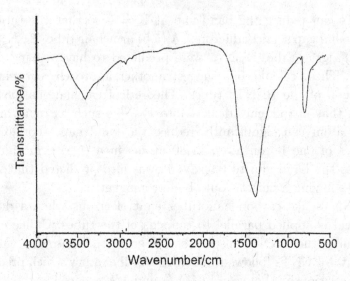

Figure 5.6   Infra-red spectrum of boron nitride nanotubes. (Ref.11)

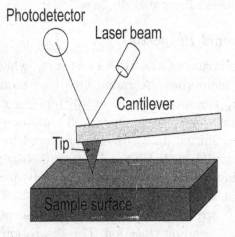

Figure 5.7   Illustration of an atomic force microscope.

curvature in the order of nanometers at its end. The tip is mounted on the end of a cantilever typically made of silicon. When the tip is brought close to the surface of a material forces between the tip and the surface produce a deflection of the cantilever. These forces

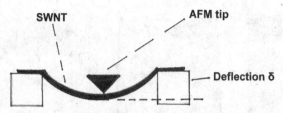

Figure 5.8   Illustration of use of an atomic force microscope to measure the strength of a nanotube deposited over a pore.

can be Van der Waals forces, electrostatic forces and chemical bonding interactions. The deflection of the cantilever is measured by the reflection of a laser beam from the top surface of the cantilever as shown in Figure 5.7. The AFM can be used to measure the mechanical properties of a single nanotube by depositing the nanotube over a nanosized pore and using the tip of the AFM to apply a load to the tube which is effectively a nanotube beam. The deflection of the tube, δ is then measured as a function of the applied force as illustrate in Figure 5.8. This method has been used to measure the mechanical strength of BNNTs.[12] The measurements were made for BNNTs of different thicknesses and it was found that the thinner tubes were stronger. For example the tensile stress of a 0.07 nm thick tube was 61.1 ± 9.2 GPa while that of a 0.34 nm tube was 13.6 ± 2.1 GPa. These values are in the same range as those of SWNTs. This means that BNNTs can be used to reinforce and increase the strength of polymers and other materials. For example it has been shown that Youngs modulus of epoxy can be increased by 66% with the addition of 1 volume % of BNNTs.[13] It was also observed that the epoxy-BNNT composites had lower electrical resistance and higher thermal conductivity.

5.2 Boron Nanotubes

5.2.1 *Structure*

There is no experimental determination of the structure of single walled boron nanotubes. However, many different structures have been proposed. Figure 5.9 illustrates three examples.[14] Figure 5.10a

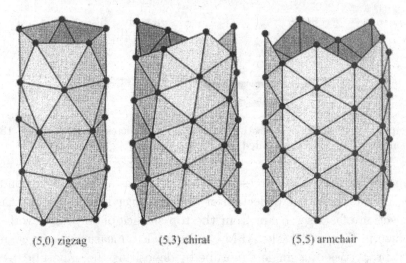

(5,0) zigzag   (5,3) chiral   (5,5) armchair

Figure 5.9   Some examples of proposed structures of boron nanotubes.

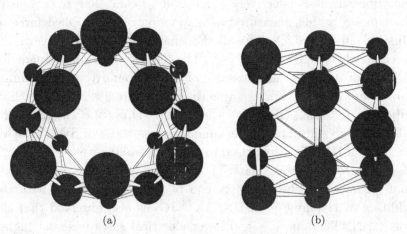

(a)   (b)

Figure 5.10   (a) The end view of a (0,6) boron nanotube obtained by minimizing the energy using density functional theory. (b) a segment of the side view of the tube. (Ref.15)

is the end view of an (0,6) armchair boron nanotube tube obtained by minimizing the energy employing density functional theory.[15] Note that the tube is not cylindrical as the shape at the end is some-what triangular. Figure 5.10b shows a segment of the side view of

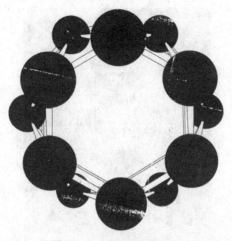

Figure 5.11   A proposed structure of a (0,6) boron nanotube which was shown not to be at the minimum on the potential energy surface. (Ref.15)

the tube. Calculation of the frequencies of this structure yielded no imaginary frequencies indicating that the structure is at a minimum on the potential energy surface. Many of the proposed structures when optimized using density functional theory yielded imaginary frequencies. As an example Figure 5.11 shows a proposed structure for the (0,6) armchair tube.[16] However, when optimized by density functional theory it yielded imaginary frequencies. The emphasis has been on armchair boron tubes because it has been argued that that zigzag tubes are not stable. Clearly much more research is needed before a definitive structure can be determined for single walled boron nanotubes.

5.2.2 *Synthesis*

The first synthesis of a single walled boron nanotube was accomplished by reacting BCl_3 with H_2 over an Mg template having cylindrical pores of 3.6 nm diameter.[17] The Mg template sitting on an alumina plug was placed in a furnace and heated under a continuous flow of H_2 to 870°C. Then BCl_3 was added to the H_2 flow for 45 minutes. Transmission electron microscope images of the recovered

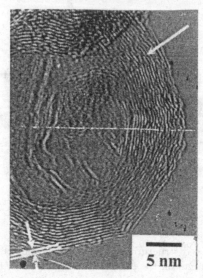

Figure 5.12   A high resolution transmission electron microscope image of boron nanotubes. (Ref.19)

material indicated boron tubes of approximately 3 nm diameter suggesting that the tubes grew in the pores of the template.

Boron nanotubes have also been synthesized by heating mixtures of boron powder and B_2O_3 in the presence of 10 nm nanoparticles of Fe_2O_3.[18] The mixtures were heated to over 1000°C in the presence of flowing argon for over 2 hours. A dark brown black material was produced on the substrate. Figure 5.12 shows a high resolution transmission electron microscope image of the material showing the presence of boron tubes.[19]

5.2.3 Electronic Structure

Since the electronic structure of a material depends on the arrangement of the atoms in the structure which is not definitely determined for a single walled boron nanotube, a definitive determination of the electronic structure is not possible. Figure 5.13 shows the results of a calculation of the energy levels of the HOMO and the LUMO versus the K vector for the structure shown in Figure 5.10.[15] Figure 5.14 shows a calculation of the density of states for the same structure.[15]

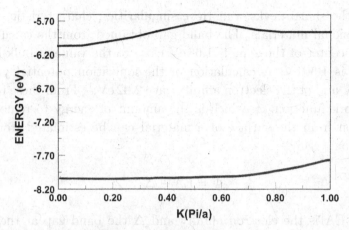

Figure 5.13 A density functional calculation of dependence of the HOMO and LUMO energy levels on the K vector for a boron tube using the structure in Figure 5.10 as the unit cell. (Ref.15)

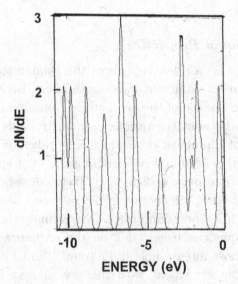

Figure 5.14 A calculation of the density of states for the structure in Figure 5.10. (Ref. 15)

Note the sharp peaks, van Hove singularities, characteristic of low dimensional materials. The band gap obtained from the calculation at the center of the zone is 1.68 eV close to the value of bulk boron which is 1.56 eV. A calculation of the ionization potential yielded 7.35 eV and of the electron affinity gave 2.02 eV.[15] From these results the work function, φ, which is the amount of energy to remove an electron from the surface of a material can be estimated from the equation,

$$\Phi = \text{EA} + \Delta/2 \tag{5.1}$$

where EA is the electron affinity and Δ the band gap at the zone center.

This yields a value of 2.86 eV for the work function which means boron nanotubes should be useful in field emission devices. In fact experimental evidence has been presented that shows this is the case.[16]

5.2.4 *Vibrational Properties*

There have only been a few reports of the Raman spectra of boron nanotubes taken with different wavelength lasers. Because of this and difference in the quality of the synthetized boron nanotubes there is little agreement between the various results. Probably the most valid are the results of Cipiuparu *et al.* which were taken on a single walled boron tube using a 532 nm wavelength laser.[17] The spectra is shown in Figure 5.15. The peak at 210 cm^{-1} likely corresponds the radial breathing mode. The peaks between 300 cm^{-1} and 500 cm^{-1} resemble those observed in carbon nanotubes. Interestingly the calculation of the Raman frequencies using DFT on the structure in Figure 5.10 gives the two most intense line at 1206 cm^{-1} and 1107 cm^{-1} which are in reasonable agreement with the two highest frequency lines observed in Figure 5.15. Interpretation of the Raman spectra of single walled boron nanotubes is at a preliminary stage and requires further work.

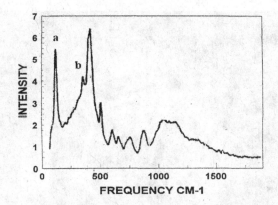

Figure 5.15   Raman spectra of boron nanotubes. (Ref. 17)

5.3 Silicon Nanotubes

5.3.1 *Structure and Motivation*

The basic switching element in computers is the metal oxide semiconductor field effect transistor (MOSFET). The MOSFET consists of thin silicon hole doped (P type) layer with an electron doped (N type) layer on each side. On top of the P layer is a thin silicon oxide layer and a metal layer which can have a voltage applied to it. If such a device could be made of silicon nanotubes, SiNTs, the device would be much faster and smaller than present MOSFETs. The smaller SiNTs would allow more switches to fit on a chip. Another possibility is that they could be used as inter-connects between present MOSFETs. Because of these and other possibilities there is much interest in producing SiNTs and determining their properties. A number of theoretical calculations have shown that SiNTs should be stable and have predicted their structure. Figure 5.16 shows an example of a structure of a (3,3) armchair SiNT obtained by a calculation.[18] Notice the hexagonal shape of the tube.

5.3.2 *Fabrication*

While silicon nanowires have been fabricated, the synthesis of SiNTs has proved to be challenging. The first successful synthesis was

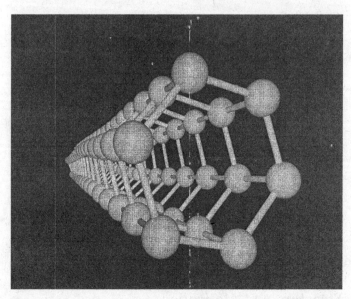

Figure 5.16   Example of a calculated structure of an armchair silicon nanotube.

achieved by a chemical vapor deposition process similar to that used to make carbon nanotubes described in Chapter 4 using a furnace similar to that shown in Figure 4.2.[19] An Al_2O_3 layer having a thin layer of gold on it and containing nanosized pores is placed in the quartz tube inside the tubular oven. The tube is evacuated and heated to 620°C at which temperature, a mixture of argon, hydrogen and silane gas are flowed into the tube. The pressure in the tube is held at 1450 Pa and the temperature at 620°C. After the process the substrate is removed and the deposited material removed from it using a dilute solution of HCl. This solution is placed on a copper grid so the material can be analyzed by transmission electron microscopy (TEM). Figure 5.17 shows the TEM image of a silicon nanotube having an approximate diameter of 44 nanometers.[19]

Later SiNTs were synthesized by molecular beam epitaxy.[20] In this method a beam of silicon atoms is incident on a substrate in a chamber having a high vacuum. The substrate consists of an alumina template having nanosized pores.

Figure 5.17   A TEM image of a silicon nanotube synthesized by a chemical vapor deposition process. (Ref. 19)

The SiNTs grew on the edges of the pores. No catalyst was used in the process. A Raman line was observed at $486\,\mathrm{cm}^{-1}$ which was attributed to amorphous silicon. The line shifted up to $521\,\mathrm{cm}^{-1}$ when the tubes were subjected to oxygen annealing due to the formation of SiO_2 on the surface of the SiNTs. No vibrations characteristic of tubes such as radial breathing modes or G and D modes were observed as in carbon nanotubes. The Raman spectra did not provide evidence for the formation of SiNTs .Photoluminescence measurements of the tubes showed a broad peak at $2.06\,\mathrm{eV}$. and a smaller peak at $3.10\,\mathrm{eV}$. The former shifted to higher values when the tubes were subjected to oxygen annealing. Considerably more research is needed to experimentally determine the properties of the SiNTs.

5.3.3 *Electronic Properties*

There have been no definitive experimental determinations of the electronic properties of SiNTs. However, there have been a number of theoretical calculations employing density functional theory which indicate SiNTs are stable.

The calculations also provide some indication of the electronic properties. For example the band gaps of single walled armchair and zigzag SiNTs have been calculated as a function of the diameter of the tubes.[20] The calculations were performed on tubes having no hydrogen atoms bonded to them and tubes with H atoms bonded to the surface of the SiNTs. Calculations of the band gap of non hydrogenated (n,0) and (n,n) silicon tubes versus diameter indicated the gap was independent of diameter above about 13Å having a value of 2.5 eV.

References

1. N. G. Chorra, Science, *269*, 996 (1995)
2. M. W. Smith *et al.* Nanotechnology, *20*, 505604 (2009)
3. D. P. Yu *et al.* Applied Physics Lett. *72*, 1966 (1988)
4. X. Blasé et al. Europhys Lett. *28*, 335 (1994)
5. T. Oku, N. Koi, and K. Suganuma, J. Phys. Chem. Solids, *69*, 1228 (2008)
6. K. H. Khoo, M. S. C. Mazzoni and S. G. Louie, Phys. Rev. *B69*, 201401 (2004)
7. M. Miller and F. J. Owens, Solid State Comm. *151*,1001 (2011)
8. X. H. S. Kang, J. Phys. Chem. *110*, 4621 (2006)
9. J. Cumings and A. Zettl, Solid State Comm. *129*, 661 (2004)
10. R. Arenal *et al.* Nano Letters *6*,1821 (2006)
11. P. Cai *et al.* Solid State Comm. *133*, 621 (2005)
12. R. Arenal *et al.* Nanotechnology, *22*, 265704 (2011)
13. H. Yan, Y. Tang, J. Su and X. Yang, Applied Physics, *A114*, 331 (2014)
14. R. K. F. Lee, B. J. Cox and J. M. Hill, Nanoscale, *2*, 859 (2010)
15. M. Miller and F. J. Owens, unpublished
16. K. C. Lau, R. Orlando and R. Pandy, J. Phys. Condens Matter, *20*, 125202 (2008)
17. D. Ciuparu, R. F. Klie, Y. Zhu, and L. Pfefferle, J. Phys. Chem. *B108*, 3967 (2004)
18. J. Bai, X. C. Zeng, H. Tanaka and J. Y. Zeng, Proc. Natl. Acad. Sci. *101*, 2664 (2004)
19. J. Sha *et al.* Adv. Mater. *14*, 1219 (2002)
20. G. Seifert, Phys. Rev. *B63*, 193409 (2001)

Chapter 6

Graphene

6.1 Structure and Fabrication

Graphene is a two dimensional array of carbon atoms where the carbon atoms have the same arrangement as the atoms in the planes of graphite. Its structure is illustrated in Figure 6.1. It had been predicted that two dimensional solids such as graphene were thermodynamically unstable and would likely distort from planarity.[1,2] The invalidity of this idea was demonstrated when scientists in England and Russia produced single graphene sheets. The discovery generated much research activity in the materials science community because the unique electronic properties of graphene suggested the potential for many applications. Properties such as its high strength and high electron mobility in the order of a million times more the copper, suggested applications such as much higher speed field effect transistors and enhancement of the strength of composites.

The first fabrication of graphene was quite simple.[3] Scotch tape was pressed on the large surface of single crystals of graphite. The graphite planes are parallel to the large facet of the crystal. The tape was carefully pealed off and then gently rubbed onto a silicon oxide substrate. The flakes on the substrate were of varying thicknesses. The thicknesses were determined by examining the flakes in a microscope using white light which consists of many wavelengths. Because the thicknesses of the flakes was in the order of angstroms, there was interference of the different wavelengths reflected from the front and

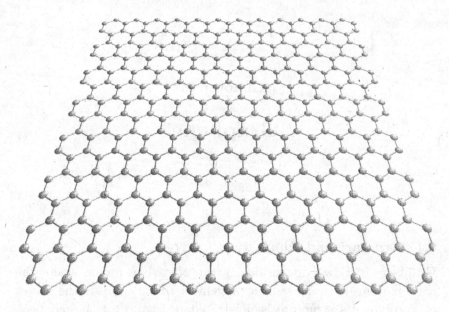

Figure 6.1   Illustration of the structure of a graphene sheet.

the back surfaces of the flakes giving the differences in color shown in Figure 6.2.[4] This enabled an identification of those flakes which were one atom thick. The upper image in Figure 6.2 shows a transmission electron microscope image of a one atom thick single crystal of graphene on an array of gold wires. Notice that the crystals are transparent. The lower image in the figure is an optical microscope image of flakes of different thicknesses indicated by the different colors. While the mechanical exfoliation method (the tape method) can provide small laboratory samples that allow measurements of properties, the development of applications requires chemical methods to produce large quantities.

One method that has been tried is to exfoliate graphite oxide (GO) in aqueous solution followed by removal of the oxygen atoms (reduction) using hydrazine.[5]

Graphite oxide can be made by a process developed by Hummers and Offeman.[6] Exfoliation, which means increasing the distance between the planes of GO, is achieved by suspending the GO in

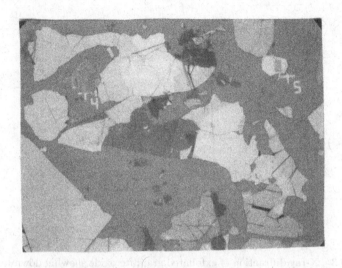

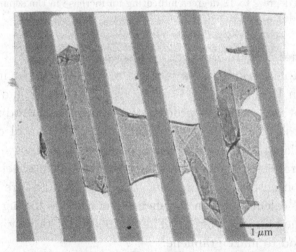

Figure 6.2 The lower image is an optical microscope image of flakes obtained by peeling off layers from a single crystal of graphite using scotch tape. The different colors indicate layers of different thicknesses. The upper image is that of a single layer taken with a transmission electron microscope. (adapted from Ref. 4)

distilled water and applying sonication which is the application of ultrasound. Figure 6.3 shows the x-ray diffraction of GO subjected to this process, showing a downward shift of the (001) reflection from a value of 2θ of $26°$ to $12°$ indicating a considerable separation

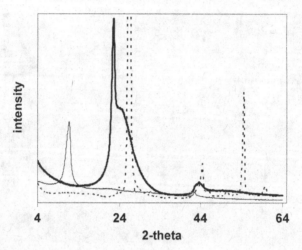

Figure 6.3   X-ray diffraction of exfoliated graphite oxide showing downward shift of line at 26 degrees to 12 degrees indicating an increase in the separation of the graphite oxide layers. (Ref. 7)

of the planes of GO.[7] However, as discussed in Chapter 3 the oxidation of graphite results in an increase of the ratio of the intensity of the Raman D mode to the G mode (see Figures 3.2 and 3.3) indicating that the oxidation process of graphite has introduced many defects into the structure.[7] This suggests that removal of oxygens from exfoliated GO to form graphene layers would leave a heavily defected graphene. It has been shown that graphene sheets having a vacancy concentration greater tha 10^{-3} carbon atoms are unstable.[8] Also DFT calculations on graphene nanoribbons having vacancies indicate they are not planar.[9] Thus exfoliating GO may not be the best method to make graphene.

Single atom thick graphene layers have been grown on the surface of silicon carbide crystals.[10] The SiC was first etched using hydrogen. It was then heated to 1200°C in a high vacuum chamber. This process evaporates the Si atoms from the surface leaving a one atom thick graphite layer which has the properties of an isolated graphene sheet.

Single layers of graphene have also been grown on thin copper films by a chemical vapor deposition process using apparatus similar to that illustrated in Figure 4.2.[11] The copper films, which had been

Figure 6.4 An STM image of a single layer of graphene grown on a copper film. (Ref. 11)

immersed in acetic acid, were placed in the vacuum chamber. Then 200 SCCM of H_2 at 2 Torr was introduced into the chamber while the chamber was heated . At 1000°C 875 SCCM of CH_4 at 11 Torr was introduced and the chamber was held at this temperature for 20 minutes. Figure 6.4 shows an STM scan of graphene grown by this method on the (111) surface of a copper film.

6.2 Electronic Structure

Figure 6.5 shows the results of the calculation of the energy of the valence band and conduction band on the wave vector K.[12] The cones of energy versus wave vector in three dimensions are referred to as Dirac cones. There are two points in the Brillouin zone, K and K' where the energy at the top of the valence band is the same as the energy of the conduction band. At these points the material is

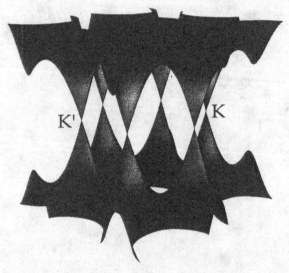

Figure 6.5   The dependence of the energy of the conduction band and valence band on the wave vector of graphene obtained theoretically. At the points K and K' the two bands are in contact. (Ref. 12)

metallic. At other points in the zone this is not the case and there is an energy gap between the conduction and valence band. Such a material is referred to as a semi-metal. Another interesting feature of the dependence of the energies on the K vector is that near the points K and K' the energy is linearly dependent on the K vector. In a semi-conductor typically the energy bands near K=0 depend quadratically on the K vector given by,

$$E = (h/2\pi)K^2/2m^* \qquad (6.1)$$

where m^* is the effective mass of the conduction electrons or holes. When the electrons or holes are subjected to an applied electric field they are accelerated with respect to the lattice as if they had a mass m^*. This mass reflects the interaction of the electrons and holes with the lattice. For a more detailed understanding of the concept of the effective mass, the reader is referred to Reference 13.

In the mid 80s theorists were examining the electronic structure of isolated graphite planes, essentially graphene, even though such a one dimensional structure did not exist. The work resulted in an unusual prediction that at the contact points between the bands, K and K' the effective mass of the conduction electrons would be zero and that they behave as though they are massless relativistic fermions.[14,15] As a result the electron dynamics in graphene is treated using the Dirac equation which is a modification of the Schrödinger equation that deals with massless relativistic particles such as photons. This means the conduction electrons move very fast in graphene in the order of 20,000 cm^2/Vs which is an order of magnitude greater than in present day silicon transistors. This makes graphene a promising candidate for electronic devices.

6.3 Anomalous Quantum Hall Effect

When a dc magnetic field is applied perpendicular to a two dimensional conductor, whose energy is quadratically dependent on the wave vector, the electrons move in circular orbits about the dc magnetic field. Each orbit has an energy given by,

$$E = (h/2\pi)\omega_c(N + 1/2) \tag{6.2}$$

where ω_c is the frequency of rotation of the electrons referred to as the cyclotron frequency. The energy levels are separated by $(h/2\pi)\omega_c$ and are known as Landau levels. The density of states is shown on the left side of Figure 6.6 showing equally spaced sharp peaks. This occurs because of the wave nature of the electrons which can only exist in specific orbits such that the wavelengths are integer multiples of the circumference of the orbits. However for metals at room temperature the spectrum on the left side of Figure 6.6 is not observed. This is because collisions of the electrons with defects, impurities and the lattice vibrations disturb the orbits of the electrons. However, the quantum Hall spectrum can be observed in high quality, defect and impurity free metals at low temperature.

However, in the case of massless Dirac Fermions the energy spectrum is different and has been shown to have the form,[16]

$$E = \pm[2eB(h/2\pi)v^2(N + 1/2 \pm 1/2)]^{1/2} \tag{6.3}$$

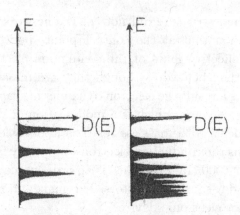

Figure 6.6   The density of states of the energy levels of the circular motion of electrons about an applied perpendicular magnetic field to a two dimensional conductor. (left). On the right are the levels for graphene showing a different distribution consistent with the conduction electrons being massless Dirac Fermions. (Ref. 12)

where v is the velocity of the electrons and N is a quantum number having values $1, 2, 3 \ldots$. This yields a very different separation of energy levels shown on the right side of Figure 6.6. This is referred to as the anomalous quantum hall effect and has been observed in graphene.[17,18] This observation provides direct evidence that the conduction electrons in graphene behave as massless Dirac fermions.

6.4 Vibrational Properties

The top spectrum in Figure 6.7 is the Raman spectrum of a single atomic layer of graphene taken using a 633 nm laser.[19] The spectrum is similar to that of graphite shown in Figure 3.2 and the vibrations have similar origins but different frequencies. The 1578 cm^{-1} line is the G mode which is due to a C-C stretching mode in the plane of the graphene. The 1347 cm^{-1} line is the D mode due to a breathing of the sp^2 C-C bonds. The line at 2686 cm^{-1} is an overtone of the D mode called the 2D mode. The frequency of this mode has been shown to depend on the number of graphene layers in the sample.[20] The second and third spectra from the top in Figure 6.7 correspond to a two and three layered sample of graphene. As the number of

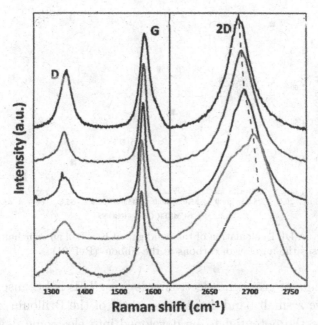

Figure 6.7 Raman spectrum of a single atomic layer of graphene (top). The lower spectra corresponds to an increasing number of layers. (Ref. 19)

layers increases the 2D spectrum shifts to higher frequencies and a shoulder emerges on the low frequency side. The bottom spectrum shows the 2D spectrum of graphite. The G spectrum shows much smaller downward shifts with increasing number of layers. So Raman spectroscopy is a powerful tool to determine the presence of a single layer of graphene.

The origine of the low frequency shoulder on the 2D band is unclear. It may be a result of the interaction between the graphene planes when there is more than one plane present. The interaction between the planes causes a slight splitting of the π and π^* bands which may be coupled to the C-C vibrations in the plane of the carbons.

6.5 Graphene Nanoribbons and Applications

There has been much interest in graphene nanoribbons, long narrow strips of graphene because such structures are likely to be

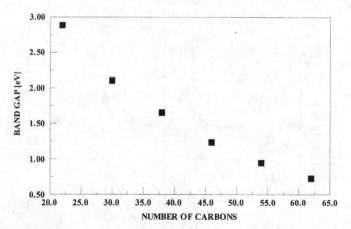

Figure 6.8 A DFT calculation of the band gap at K = 0 of an armchair graphene ribbon versus the number of carbons in the ribbon. (Ref. 21)

constituents of graphene electronic devices. This is because the ribbons have a small band gap at the center of the Brillouin zone and thus have the potential to be developed into electronic devices. As discussed above large graphene sheets are metallic at the center of Brillouin zone. However, it turns out that graphene nanoribbons have a band gap that depends on the length of the ribbon. Figure 6.8 shows the results of a DFT calculation of the band gap at the zone center of an armchair ribbon, three carbon rings wide, as a function of the number of carbons in the ribbon (effectively the length).[21] The calculation shows that as the length of the ribbon increases the band gap approaches zero. This means that relatively short ribbons will be necessary for graphene based electronic devices. Figure 6.9 illustrate one concept of a field effect transistor using graphene.[22] Two graphene nanoribbons are connected by a small chain of benzene rings. In the same plane is another graphene nanoribbon slightly separated from the other two. This ribbon serves as a gate and when a voltage is applied to it, current flows across the bridge between the two other ribbons.

Developing graphene electronic devices based on graphene ribbons will likely require ribbons of 10 nm or less. However, their large scale fabrication will have to produce ribbons that are defect free

coplanar graphene side gate

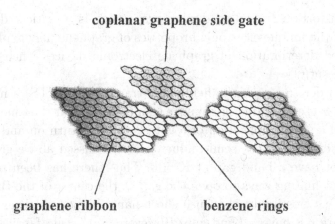

graphene ribbon　　　　**benzene rings**

Figure 6.9　A concept of a field effect transistor using graphene ribbons. (Ref. 22)

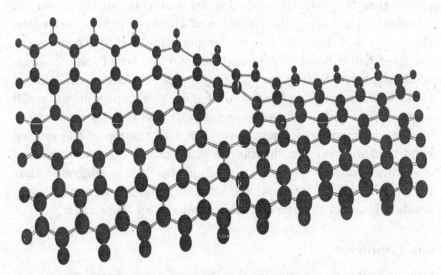

Figure 6.10　A DFT calculation of the minimum energy structure of an armchair graphene nanoribbon having one carbon vacancy. (Ref. 9)

which may be a challenge. DFT calculations of the minimum energy structure of graphene ribbons having over 100 carbon atoms and containing one carbon vacancy indicate they are significantly distorted from two dimensions.[9] Figure 6.10 shows the calculated minimum energy structure of an armchair ribbon with a carbon vacancy.

This distortion from two dimensions, which is a critical determinant of the unique electronic properties of graphene may explain the observed deterioration of graphene electronic devices when defects are present.[23]

In order to be used in field effect transistors (FETS) a material has to be a semiconductor having a small band gap at the zone center. This allows the application of a gate voltage to turn on and off the flow of current in the semiconductor. As discussed above graphene does not have a band gap at $K = 0$. Thus there has been research aimed at finding ways to open the gap at the center of the Brillouin zone. One possibility is to use short nanoribbons which as shown in Figure 6.8 have a band gap. However, large scale fabrication of such short narrow ribbons could be a challenge. Recently it has been shown that by thermally decomposing a silicon carbide substrate, a carbon layer having the structure of graphene can be formed on the surface and depending on the temperature used to decompose it, the layer had a band gap centered around the Fermi level.[24] Angle resolved photo electron spectroscopy measurements of SiC heated at 1360°C indicated a band gap of 0.5 eV. Samples grown at 20 degrees lower did not have a band gap. It was also shown that the effective mass and the electron velocity near the top of the valence band depended on the direction of propagation. This was the first observation of such anisotropy in graphene. It was suggested that the opening of the gap and the anisotropy resulted from periodic bonding between the graphene layer and the SiC substrate.

6.6 Graphyne

Other possible two dimensional carbon structures such as graphyne have been theoretically predicted. Graphyne consists of an array of benzene rings bonded to each other by acetylenic linear links of two carbon atoms. It was first predicted to exist in 1987 using a semi-empirical molecular orbital method (MNDO).[25] However, since then more advanced models such as DFT have been used to predict its structure. Figure 6.11 shows a predicted structure of graphyne. Figure 6.12 shows the results of calculation of the dependence

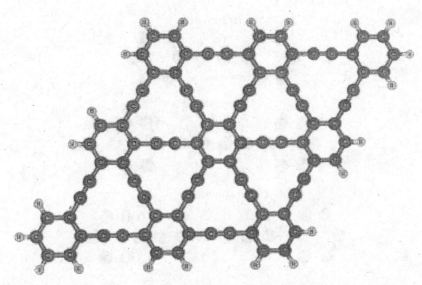

Figure 6.11   Illustration of two dimensional structure of graphyene (adopted from Ref. 25)

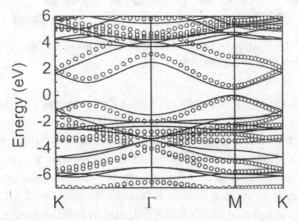

Figure 6.12   Calculation of the dependence of the energy levels of graphyne on the wave vector. (Ref. 26)

of the energy levels on the wave vector using periodic boundary condictions.[26] The interesting feature of the calculation is that at the M point in the Brillouin zone there is a direct band gap of 0.46 eV. This means that this material has the potential to be developed

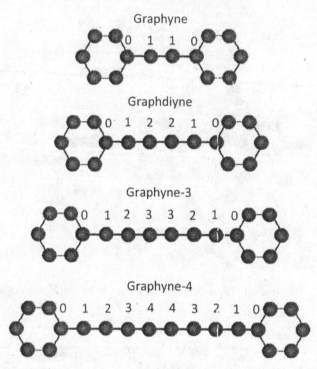

Figure 6.13 Illustration of the kinds of acetylenic links that have been predicted for the various forms of graphyne.

into electronic devices such as field effect transistors. It also has been predicted that there are other forms of this material which have longer acetylenic links having more than two carbon atoms. Figure 6.13 shows some of the other linkages that are possible and their designations. Calculations of the mechanical properties show that graphyne is strong having a tensile strength 81.2 GPa but not as strong as graphene which has a calculated tensile strength of 119.2 GPa.[27] While graphyne is an interesting material, its synthesis has proven a challenge. There have been no reports of synthesis of sufficient quantities of the material to allow experimental determination of properties. Thus to date properties have been obtained by theoretical methods.

References

1. R. E. Peierls, Ann. I. H. Poincare *5*, 177 (1953)
2. L. D. Landau, Phys. Z. Sovjetunion, *11*, 26 (1937)
3. K. S. Novoselov *et al.* Science, *306*, 666 (2004)
4. J. C. Myer *et al.* Nature *446*, 60 (2007)
5. S. Stankovich *et al.* Nature, *442*, 282 (2006)
6. W. Hummers and R. Offeman, J. Am. Soc. *80*, 1339 (1958)
7. F. J. Owens, Mol. Phys. *113*, 1280 (2015)
8. R. R. Nair *et al.* Nature Physics, *8*, 199 (2012)
9. M. Miller and F. J. Owens, Chem. Phys. Lett. *570*, 42 (2013)
10. C. Berger *et al.* J. Phys. Chem. *B108*, 19912 (2004)
11. G. W. Flynn. J. Chem. Phys. *135*, 050901 (2011)
12. M.I. Katsnelson, Materials Today, *10*, 20 (2007)
13. C. Kittel, Introduction to Solid State Physics, 6th Edition, 1986, p194 John Wiley and Sons Inc., New York
14. G. W. Semenoff, Phys. Rev. Lett. *53*, 2449 (1984)
15. F. D. Haldane, Phys. Rev. Lett . *61*, 2015 (1988)
16. V. P. Gusynin and S. G. Sharapov, Phys. Rev. *B71*, 125124 (2005)
17. K. S. Novoselov *et al.* Nature, *438*, 197 (2005)
18. Y. Zhang *et al.* Nature, *438*, 201 (2005)
19. F. J. Owens (unpublished)
20. A. C. Ferrari, Solid State Comm. *143*, 47 (2007)
21. F. J. Owens, J. Chem. Phys. *128*, 194701 (2008)
22. B. Ozyilmaz *et al.* Phys. Rev .Lett. *99*, 166804 (2007)
23. X. Xuan *et al.* Appl. Phys. Lett. *92*, 013101 (2008)
24. M. S. Nevius *et al.* Phys. Rev. Lett. *115*, 136802 (1915)
25. R. H. Baughman, H. Eckhard, and M. Kertesz, J. Chem. Phys. *87*, 6687 (1987)
26. J. Kang *et al.* J. Phys. Chem. *C115*, 20466 (2011)
27. T. Shao *et al.* J. Chem. Phys. *137*, 194901 (2012)

Chapter 7

Other Low Dimensional Materials

In this chapter the electronic properties of other low dimensional materials will be discussed. The emphasis will be on materials for which there is experimental evidence for their existence.

7.1 Silicene

The fabrication of graphene and the determination of its unique electronic properties such as massless conduction electrons raised the question are there other two dimensional materials which could have similar properties?

Since silicon is in the same column in the periodic table as carbon and thus has the same s^2p^2 valence structure, it may be that silicon could form a similar structure as graphene, which has been named silicene, and have analogous unique electronic properties. Silicene could have the potential for interesting applications because of its chemical compatibility with present day electronic devices such as field effect transistors, which are largely made of silicon.

One atom thick silicon sheets have been fabricated on the (111) surface of silver.[1] Clean Ag surfaces were first prepared by bombardment with Ar^+ ions and subsequent annealing of the Ag at 530°C for 30 minutes in an ultra high vacuum. Silicon was then deposited on the Ag surface by heating a sample of Si in the vacuum chamber while the Ag surface was held at a temperature of 220°C.

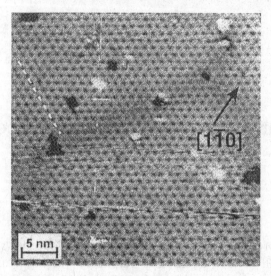

Figure 7.1   An STM image of a one atom thick sheet of silicon deposited on the (111) surface of silver. (Ref. 1)

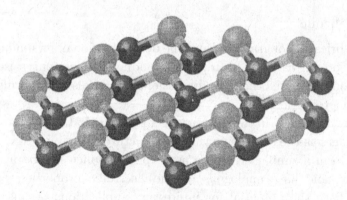

Figure 7.2   The puckered structure of silicene. (Ref. 3)

Figure 7.1 shows an STM image of the 2D silicon on the Ag surface. Molecular orbital calculations of free standing 2D silicon ribbons and sheets predict two stable structures.[2,3,4,5] One structure is the same as that of graphene and is planar as illustrated in Figure 6.1. The other structure is slightly puckered. This structure is illustrated in Figure 7.2. The structure consists of two parallel planes of silicon atoms identified by the dark and light shaded atoms in the figure. In each plane the atoms form an hexagonal arrangement. The planes are bonded to each other and are separated by 0.53 A. The calculations predict that the buckled structure is slightly more stable than the flat structure.[3,4] Both structures have similar electronic properties as graphene.[2,3,4] In both the π^* and π energy levels, the valence and conduction bands, touch at points in the Brillouin zone and near these points the dependence on K is linear. Consequently as in graphene the electron charge carriers behave as massless relativistic Fermions. In the case of planar 2D silicon nanoribbons, it was predicted using DFT that only armchair ribbons were stable.[5] The calculation predicted zig-zag ribbons to have imaginary frequencies meaning they are not at a minimum on the potential energy surface.

7.2 Boron Nitride

There are other three dimensional solids that have layered structures such as boron nitride. Its structure, which is similar to graphite, is illustrated in Figure 7.3. The dark circles are nitrogen atoms and lighter shaded circles are boron atoms. As in graphite the planes are bonded to each other by Van der Waals potentials which are considerably weaker than the covalent bonds between the boron and nitrogen atoms in the planes. Because of this layered structure and weaker bonding between the planes, it was possible to exfoliate the BN crystal into flakes some of which were one atom thick.[6] A fresh surface of the BN crystal was rubbed on the surface of another crystal such as silicon oxide. The flakes were then examined by the using an optical microscope and the single atom thick flakes were identified, as described in Chapter 6 for graphene. Atomic force microscopy was also used to determine the thickness. Measurement of the electrical conductivity of the one atom thick flakes indicated they were insulating. DFT calculations of armchair and zig-zag boron nitride ribbons confirm they have large band gaps.[7,8] Figure 7.4(a) and (b) shows the minimum energy structure of an armchair and zig-zag BN nanoribbon. The calculations indicated these were minimum energy structures because they yielded no imaginary frequencies. Figure 7.5(a)

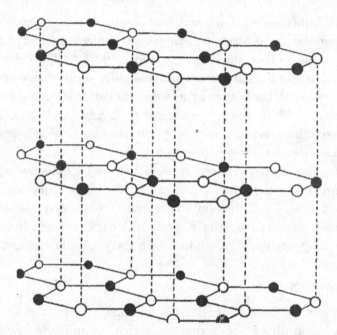

Figure 7.3   Crystal structure of hexagonal boron nitride showing its layered structure.

shows the calculated dependence of the band gap at the center of the Brillouin zone on the length of an armchair BN nanoribbon showing it is not strongly dependent on ribbon length. Figure 7.5(b) shows the same dependency for an zig-zag ribbon indicating an oscillatory dependence of the band gap on ribbon length. Figure 7.6 is the calculated dependence of the energy of the conduction band and valence band for a zig-zag BN ribbon on the wave vector K showing the band gap remains large for all values of K.[7] A similar result is obtained for an armchair ribbon. The large band gaps of the BN two dimensional structure seem to suggest that BN sheets do not have the application potential of graphene in electronic devices. However, calculations have predicted that changing the ratio of B to N in the structures can significantly lower the band gap to where the structures could be semiconducting.[8] Calculations have also shown that applying an electric field parallel to the long direction of the ribbon can significantly reduce the magnitude of the band gap. In fact, it is shown

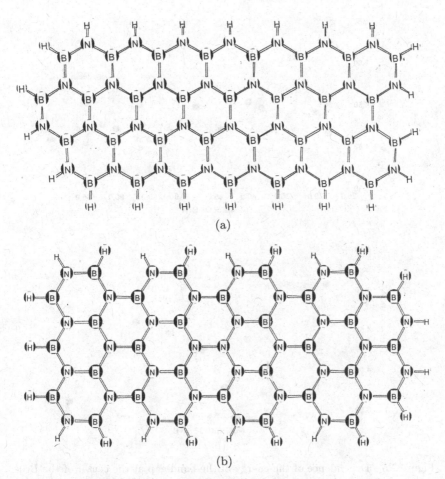

Figure 7.4 Structure of an armchair BN nanoribbon (a) and a zig-zag ribbon (b) obtained by a DFT calculation. (Ref. 8)

that it is possible to reduce it to zero making the ribbons metallic and conducting.[9,10] Figure 7.7 is a plot of the magnitude of the band gap versus the strength of the electric field for an armchair ribbon of a 20Å length. The calculations predict that the decrease of the band gap depends on the length of the ribbon with it decreasing more rapidly for longer ribbons. This opens up the possibility of using an electric field to switch the conductivity on and off which is effectively what happens in a field effect transistor.

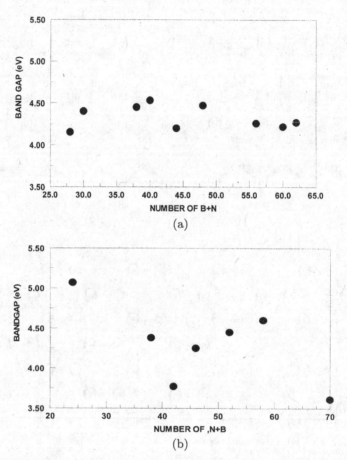

Figure 7.5   Dependence of the energy of the band gap at the center of the Brillouin zone versus the length of the ribbon for an armchair BN ribbon (a) and a zig-zag ribbon (b) obtained by DFT. (Ref. 8)

7.3 Dichalcogenides

Dichalcognides refer to compounds of the form MX_2 where M is a transition metal such as Mo, Nb or V and X is typically sulpher or selenium. These materials are layered structures and some of them can be exfoliated into single layered materials because of the weak Van der Waals interactions between the layers. Figure 7.8 is an illustration of the crystal structure of MoS_2. This structure is identified

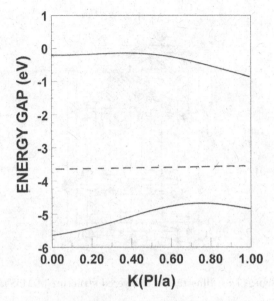

Figure 7.6   Dependence of the energies of the conduction and valence levels on the wave vector K for BN nanoribbons calculated by DFT. (Ref. 7)

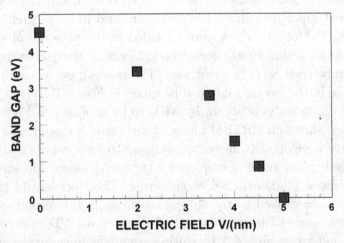

Figure 7.7   Plot of the dependence of the energy of the band gap on the strength of an applied electric field for an armchair BN nanoribbon. (Ref. 9)

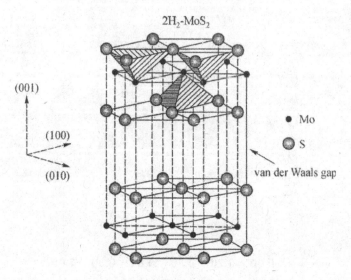

Figure 7.8   Illustration of layered structure of MoS$_2$.

as an H structure having D$_{6h}$ symmetry. Figure 7.9 shows the structure of NbSe$_2$ which has D$_{3d}$ symmetry and is referred to as the T structure. Single layers of these materials have been fabricated by the same methods used to make BN sheets described in Section 7.2.[6] Chemical vapor deposition has also been used to make single layers of MoS$_2$.[11] MoS$_2$ powders were annealed in the presence of sulpher vapor in an argon atmosphere. Single layers of MoS$_2$ grew between two sandwiched SiO$_2$-Si substrates. The layers grew as small flakes as shown in the electron microscope image in Figure 7.10 (top). Over 70% of the flakes were shown by AFM to be monolayers. Figure 7.10 (bottom) shows an HRTEM image of the flakes. Subsequently methods were developed to more precisely control the number of layers produced.[12] The method employed a two step process. Mo was sputtered onto a (100) oriented Si substrate. The thickness of the Mo layers was controlled by the sputtering times. In the second step the Mo films were subjected to sulphur vapor in a CVD apparatus at 600°C and a pressure of 5 Torr using argon gas to carry the sulpher vapor to the Mo films. The thickness of the MoS$_2$ films was determined by AFM, TEM and Raman spectroscopy.

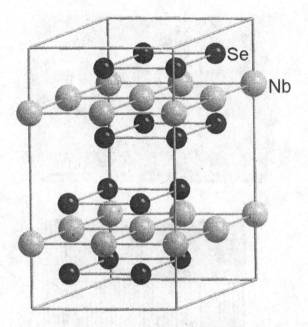

Figure 7.9   Illustration of layered structure of NbSe$_2$.

The Raman spectra of MoS$_2$ using an 488 nm laser consists of a line at 383 cm^{-1} which is due to an in plane E$_{2g}$ Mo-S vibration and a line at 404 cm^{-1} which is A$_{1g}$ out of plane vibration. The A$_{1g}$ mode was shown to shift to higher frequencies as the sample thickness increased as measured by AFM.[12] Figure 7.11 is a plot of the of the frequency of the A$_{1g}$ mode versus thickness showing that Raman spectroscopy can be used to measure the thickness of the films.

One layer thick flakes of MoS$_2$ were connected between gold electrodes which were on top of an oxidized Si layer.[6] A voltage was applied to the SiO$_2$ layer which acted as a gate and the current measured. Conductivities in the range of 0 to 0.5 (MΩ)$^{-1}$ were obtained.[6] A measurement of the temperature dependence of the conductivity indicated the material was semiconducting.

A calculation of the energy levels of a single layer of MoS$_2$ yielded a band gap of 1.78 at the highest symmetry point in the zone.[13] While the discussion of layered dichalcognides has focused on one

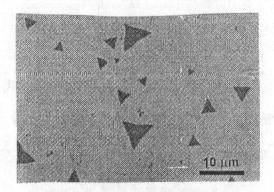

Figure 7.10   Electron microscope image of the exfoliated flakes of MoS_2 (top) and a HRTEM image of the flakes (bottom). (Ref. 11)

example, MoS_2, It should be pointed out that there are many compounds having similar structures that could be exfoliated into single layered structures, which may have interesting electronic properties with potential for device applications. Using theoretical first principles calculation of stability, 88 layered compounds of the MX_2 variety were identified.[14] The M was generally a transition metal and the X a sulphur, a selenium or an oxygen.

7.4 Black Phosphorous

Black phosphorous (BP) consists of phosphorous atoms arranged in the structure illustrated in Figure 7.12. The layers are separated by

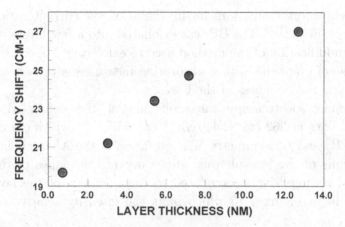

Figure 7.11 Plot of the frequency of the Raman line of the A_{1g} mode of MoS_2 versus the thickness of the material. (Ref. 12)

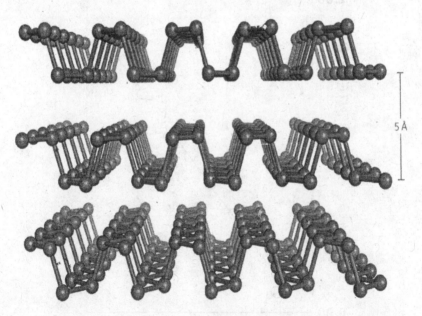

Figure 7.12 Illustration of the layered structure of black phosphorous.

5.5Å and are weakly bonded by Van der Waals potentials making it possible to produce materials of a few layers by exfoliation. Unlike graphene the material has a band gap at the zone center typical of a semiconductor. The material having a few layers has been made into field effect transistors having large on–off current ratios and high mobilities.[15,16] The BP was exfoliated into a few layers using a slight modification of the method used to exfoliate graphene.[17] As in the case of graphene optical absorption measurements were used to determine the thickness of the flakes.

Raman spectroscopy measurements of the exfoliated flakes yielded lines at 362 cm^{-1}, 440 cm^{-1} and 467 cm^{-1} which correspond to A_g^1, B_g^2 and A_g^2 symmetry. The vibration of the A_g^1 mode involves vibrations of the phosphorous atoms out of the planes of the layers and its intensity is correlated to the thickness of the layers as shown in in Figure 7.13 which plots the intensity relative to a Si

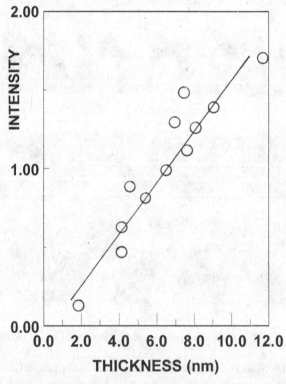

Figure 7.13   The intensity of the A_{1g} Raman line of black phosphorous versus thickness of the flakes. (Ref. 17)

standard versus the thickness of the layers. Photo luminescence measurements yielded an optical photon gap of 1.6 eV for a layer of 1.6 nm thickness.

One concept of FETS based on using 2D materials employs nanoribbons as illustrated in Figure 6.9. Because of the anisotropic structure of the planes of phospherene shown in Figure 7.12, the electronic structure of phospherene nanoribbons depends on the direction in the planes from which they are formed.

This is demonstrated by calculations of the electronic structure of the ribbons formed in different directions in the planes.[18] Figure 7.14 shows the unit cell of phospherene used to calculate the dependence of the HOMO and LUMO energy levels on the wave vector. Figure 7.15a shows the dependence of the energy levels on the K vector for ribbons formed parallel to the line between atoms 7 and 6 in Figure 7.14. Figure 7.15b shows the dependence for ribbons formed perpendicular to this direction, showing a larger band gap at the zone center compared to that in Figure 7.15a

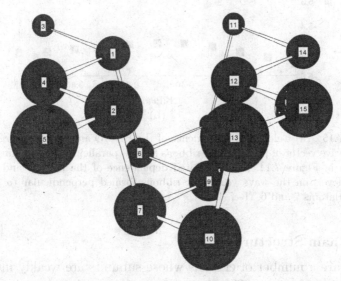

Figure 7.14 Unit cell used to calculate dependence of energy levels on wave vector of phospherene nanoribbbons.

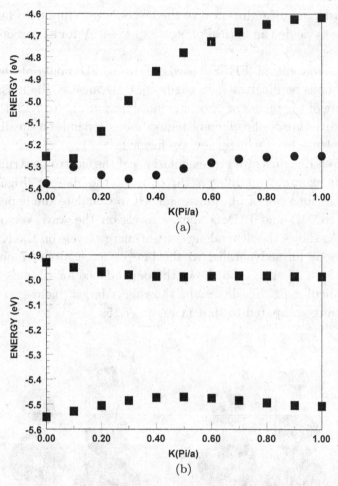

Figure 7.15　(a) Calculated dependence of the HOMO and LUMO energy levels on the wave vector of phospherene ribbons formed parallel to a line joining atoms 7 and 6 in Figure 7.14. (b) Calculated dependence of the HOMO and LUMO energy levels on the wave vector for ribbons formed perpendicular to the line between atoms 7 and 6. (Ref. 18)

7.5　Chain Structures

There are a number of crystals whose subunits are weakly interacting chains of atoms. The electronic structure of these crystals can be approximated by the electronic structure of a single chain. In

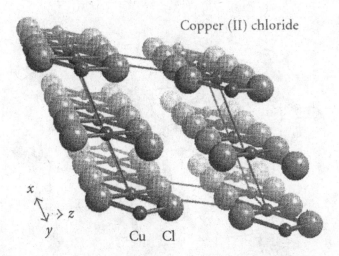

Copper (II) chloride

Figure 7.16   Crystal structure of copper chloride CuCl$_2$.

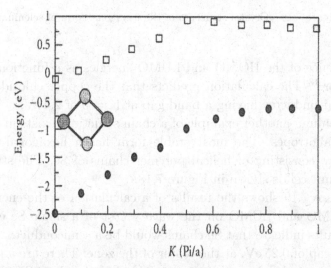

Figure 7.17   Results of a calculation of the energies of the HOMO and LUMO levels of a single chain of copper chloride versus the K vector. (Ref. 19)

principle it should be possible to exfoliate these crystals into layered structures which consist of parallel chains. However, this has not been reported. Figure 7.16 shows the crystal structure of copper chloride, one example of this kind of crystal. Figure 7.17 is a calculation of the

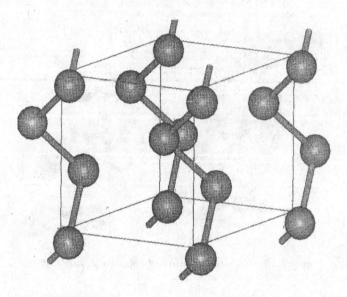

Figure 7.18   Crystal structure of the hexagonal form of selenium.

dependence of the HOMO and LUMO energies as a function of the K vector.[18] The calculation predicts that the copper chloride chain is a semiconductor having a band gap at K = 0 of 2.4 eV.

Selenium, another example of a chain structure, exists in a number of allotropes. The most stable form has a hexagonal crystal structure consisting of helical polymer chains of Se. The structure of the unit cell is shown in Figure 7.18.

Figure 7.19 shows the results of a calculation of the energies of the HOMO and LUMO on the wave vector of a single Se chain.[20] The results indicate that Se chains would be a semiconductor having a band gap of 0.25 eV. at the center of the zone. There are a number of other crystals whose constituents are chains such as SN_x, LiCuO which has chains of CuO_2 and $CsMnCl_3$ $2H_2O$. Some of these display interesting magnetic properties which will be discussed in the next chapter.

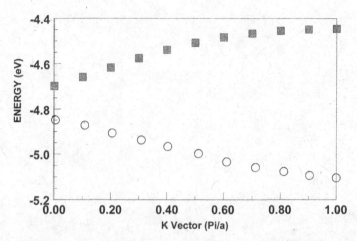

Figure 7.19 Results of a calculation of the energies of the HOMO and LUMO levels of a single chain of selenium having the structure shown in Figure 7.18 versus the K vector. (Ref. 20)

References

1. P. Vogt *et al.* Phys. Rev. Lett. *108*, 155501 (2012)
2. G. G. Guzman-Verri and L. C. L. Yan Voon, Phys. Rev. *B76*, 075131 (2007)
3. Y. Ding and J. Ni, Appl. Phys. Lett. *95*, 083115 (2009)
4. S. Chanagirov, Phys. Rev. Lett. *102*, 236804, (2009)
5. M. Miller and F. J. Owens, Chem. Phys. *381*, 1 (2011)
6. K. S. Novoselov *et al.* Proc. National Academy Science *102*, 10451 (2005)
7. A. J. Du, S. C. Smith and G. O. Lu, Chem. Phys. Lett. *447*, 181 (2007)
8. F. J. Owens, Mol. Phys. *109*, 1527 (2011)
9. Z. Zhang and W. Guo, Phys. Rev. *B77*, 075403 (2008)
10. C. Park and S. G. Louie, Nano Lett. *8*, 2200 (2008)
11. Z. Lin *et al.* APL Materials, *2*, 092514 (2014)
12. J. Park, Appl. Phys. Lett. *106*, 012104 (2015)
13. S. Lebegue and O. Eriksson, Phys. Rev.*B79*, 115049 (2009)
14. C. Ataca, H. Sahin, and S. Ciraci, J. Phys. Chem. *116*, 8983 (2012)
15. H. Liu *et al.* ACS Nano *8*, 4033 (2014)
16. S. P. Koenig *et al.* Appl. Phys. Lett. *104*, 103106
17. A. Castellanos-Gomez *et al.* 2D Materials *1*, 02501 (2014)
18. F. J. Owens, Solid State Comm. *223*, 27, (2015)
19. F. J. Owens, J. Nanoparticles, *2013*, 756473 (2013)
20. F. J. Owens, unpublished

Chapter 8

Magnetism in Low Dimensional Materials

8.1 Basics of Ferromagnetism

Ferromagnetism occurs in solids made of atoms or molecules which have a magnetic moment. Such atoms as Fe, Ni and Co have inner shells that are not filled and thus have magnetic moments.

When crystals are formed from atoms having a net magnetic moment, a number of different situations can occur depending on how the magnetic moments of the individual atoms are aligned with respect to each other. Figure 8.1 illustrates some of the possible arrangements that can occur in two dimensions. The point of the arrow is the north pole of the bar magnet associated with the atom. If the magnetic moments are randomly arranged with respect to each other, as shown in Figure 8.1a, then the crystal has a zero net magnetic moment, and this is referred to as the paramagnetic state. The application of a dc magnetic field aligns some of the moments, giving the crystal a small net moment. In a ferromagnetic crystal these moments all point in the same direction, as shown in Figure 8.1b, even when no dc magnetic field is applied. The whole crystal has a magnetic moment and behaves like a bar magnet producing a magnetic field outside of it. In an antiferromagnet the moments of identical atoms are arranged in an anti-parallel scheme, that is opposite

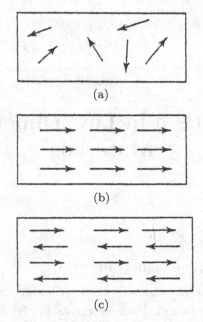

Figure 8.1   Illustration of various arrangements of individual atomic magnetic moments in 2 dimensions that constitute paramagnetic, ferromagnetic, and anti-ferromagnetic materials.

to each other, as shown in Figure 8.1c and hence the material has no net magnetic moment.

In a crystal each atom having a magnetic moment has a magnetic field about it. If the magnetic moment is large enough, the resulting large dc magnetic field can force a nearest neighbor to align in the same direction provided the interaction energy is larger than the thermal vibrational energy, kT, of the atoms in the lattice. The interaction between atomic magnetic moments is of two types, the so-called exchange interaction and the dipolar interaction. The exchange interaction is a purely quantum mechanical effect, and is generally the stronger of the two interactions.

In the case of a small particle such as an electron which has a magnetic moment $g\mu_B$, the application of a dc magnetic field forces its spin vector to align such that it can have only two projections in the direction of the dc magnetic field which are $\pm 1/2 g\mu_B$, where, μ_B, is the magnetic moment called the Bohr magneton, and g = 2.0023 is the dimensionless gyromagnetic ratio of a free electron. The wave function representing the state $+1/2g\mu_B$ is designated, α, and for $-1/2g\mu_B$ it is, β. The numbers $\pm 1/2$ are called the spin quantum numbers m_s. For a two electron system it is not possible to specify which electron is in which state. The Pauli exclusion principle does not allow two electrons in the same energy level to have the same spin quantum numbers m_s. Quantum mechanics deals with this situation by requiring that the wave function of the electrons be antisymmetric, that is it changes sign if the two electrons are interchanged. The form of the wave function that meets this condition is $(1/2)^{1/2}[\Psi_A(1)\Psi_B(2) - \Psi_A(2)\Psi_B(1)]$. The electrostatic interaction between the two electrons is given by the expression;

$$E = \int \left[\left(\frac{1}{2}\right) e^2 \middle/ r_{12} \right] [\Psi_A(1)\Psi_B(2) - \Psi_A(2)\Psi_B(1)]^2 dV_1 dV_2$$

$$(8.1)$$

which involves carrying out an integration over the volume. Expanding the square of the wavefunctions in equation (8.1) gives two terms,

$$E = \int [e^2/r_{12}][\Psi_A(1)\Psi_B(2)]^2 dV_1 dV_2$$

$$- \int [e^2/r_{12}]\Psi_A(1)\Psi_B(1)\Psi_A(2)\Psi_B(2) dV_1 dV_2 \qquad (8.2)$$

The first term is the normal Coulomb interaction between the two charged particles. The second term, called the exchange interaction, represents the difference in the Coulomb energy between two electrons with spins that are parallel and antiparallel. For an antiferromagnet it is positive and for a ferromagnetic it is negative. The exchange interaction, because it involves the overlap of orbitals, is primarily a nearest neighbor interaction, and it is generally the dominant interaction. The other interaction, which can occur in a lattice of magnetic ions, called the dipole-dipole interaction, has the form;

$$\boldsymbol{\mu}_1 \cdot \boldsymbol{\mu}_2/r^3 - 3(\boldsymbol{\mu}_1 \cdot \mathbf{r})(\boldsymbol{\mu}_2 \cdot \mathbf{r})/r^5 \qquad (8.3)$$

where **r** is a vector along the line separating the two magnetic moments μ_1 and μ_2, and r is the magnitude of this distance. The symbols in bold are vector quantities.

For ferromagnetism to exist in a crystal the density of states at the Fermi level must be different for the spin down (β) states compared to the spin up (α) states.

The magnetization M of a bulk sample is defined as the total magnetic moment per unit volume. It is the vector sum of all the magnetic moments of the magnetic atoms in the bulk sample divided by the volume of the sample. It increases strongly at the Curie temperature T_c, the temperature at which the sample becomes ferromagnetic, and it eventually becomes constant as the temperature is lowered further below T_c. It has been found empirically that below the Curie temperature the magnetization of a ferromagnet depends on the temperature as:

$$M(T) = M(0)(1 - cT^{3/2}) \tag{8.4}$$

where M(0) is the magnetization at zero degrees Kelvin, and c is a constant. This equation is known as the Bloch equation.

Generally for a bulk ferromagnetic material below the Curie temperature, the magnetization, M, is less than the magnetization the material would have if every individual atomic moment were aligned in the same direction. The reason for this is because of the existence of domains. Domains are regions in which all the atomic moments point in the same direction so that within each domain the magnetization is saturated, i.e. it attains its maximum possible value. However, the magnetization vectors of different domains in the sample are not all parallel to each other. Thus the sample has a total magnetization less than value for the complete alignment of all moments. Some examples of domain configurations are illustrated in Figure 8.2. They exist because the magnetic energy of the sample is lowered by the formation of domains. Applying a dc magnetic field can increase the magnetization of a sample. This occurs by two processes. The first process occurs in weak applied fields when the volume of the domains which are oriented along the field direction increases. The second process dominates in stronger applied magnetic fields which forces the

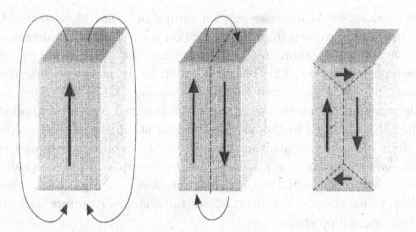

Figure 8.2   Illustrations of some examples of domain structure in ferromagnetic materials.

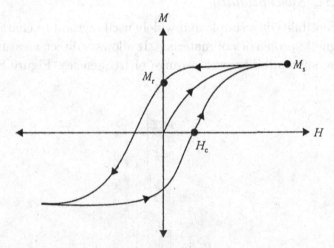

Figure 8.3   Plot of the magnetization M versus an applied magnetic field H for a hard ferromagnetic material, showing the hysteresis loop with the coercive field H_c, and the remnant magnetization, M_r.

magnetization of the domains to rotate toward the direction of the field. Figure 8.3 shows a schematic plot of the magnetization curve of a ferromagnetic material. It is a plot of the total magnetization of the sample M versus the applied dc field strength, H. Initially as H increases M increases until a saturation point M_s is reached. When H is decreased from the saturation point M does not decrease to zero magnetization, rather it has a non zero magnetization at zero magnetic field. This is called hysteresis. It occurs because the domains that were aligned with the increasing field do not return to their original orientation when the field is removed. When the applied field H is returned to the zero, the magnet still has a magnetization referred to as the remnant magnetization, M_r. In order to remove the remnant magnetization a field H_c has to be applied in the opposite direction to the initial applied field, as shown in Figure 8.3. This field, called the coercive field, causes the domains to rotate back to their original positions.

8.2 Methods of Observing Ferromagnetism

8.2.1 *AC Susceptibility*

AC susceptibility is a simple and widely used method to characterize the magnetic properties of materials. It allows a direct measurement of the ac susceptibility over a range of frequencies. Figure 8.4 is a

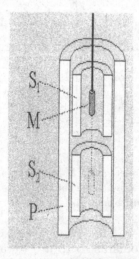

Figure 8.4   Schematic of apparatus to measure AC susceptibility consisting of a primary coil P wound around two secondary coils having equal inductances. The sample is located in one of the secondary coils labeled M. The difference in voltage between the two coils is a measure of the magnetic susceptibility.

schematic of an ac susceptibility apparatus. An ac signal is applied to a primary coil P, which is wound around two secondary coils S having an equal number of turns and thus equal inductance. One of the coils contains the sample to be characterized labeled M in the figure. The EMF induced in this coil is directly proportional to the magnetization of the sample. The difference of the ac voltage between the two secondary coils is measured, and this voltage is proportional to the susceptibility.

8.2.2 *Magnetic Resonance*

When a magnetic field is applied to an atom or molecule having an unpaired electron, the energy of the $m_s = +1/2$ is greater than the $m_s = -1/2$ and the difference depends on the strength of the applied field H as $g\beta H$ where g and β are constants. In the electron spin resonance (ESR) method radiation of a fixed frequency, ν, typically in the microwave region having energy, $h\nu$, is applied to the sample and the magnetic field slowly increased until the separation between the spin states, $g\beta H$ equals $h\nu$. When this occurs there is a transition of spin down electrons to the spin up state and microwave energy is absorbed. The situation is illustrated in Figure 8.5. The sample is in a microwave cavity, which concentrates the microwave radiation over the sample and is located between the poles of a magnet. For x-band (9.2 GHz) and $g \cong 2$, the EPR absorption occurs for fields near 0.32 Tesla. Superimposed on the sweep of the dc magnetic field is an ac modulation typically having a frequency of 100 KHz. This modulation is supplied by rf coils mounted on the side walls of the microwave cavity. The modulation of the absorption results in a time varying out-put signal at the crystal diode which changes phase by 180 degrees at the peak of the absorption. Phase sensitive detection is employed which compares the phase of this out-put with that of a reference signal, enabling the derivative of the absorption to be detected. This reduces noise and enhances the sensitivity. A block diagram of a simple electron paramagnetic resonance spectrometer is presented in Figure 8.6. The Gunn diode produces the microwaves which are transmitted via the waveguide to the microwave cavity.

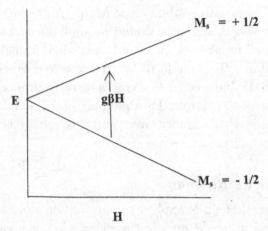

Figure 8.5   Illustration of the splitting of the m_s spin states of an unpaired electron in a dc magnetic field showing the radiation induced absorption of energy which occurs when $h\nu = g\beta H$ which is the basis of the electron spin resonance measurement.

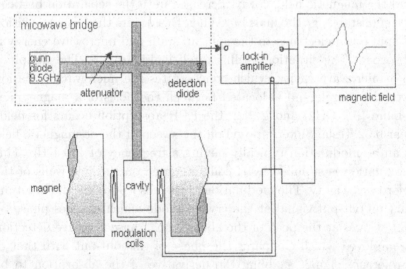

Figure 8.6   Block diagram of an ESR spectrometer. The dark areas represent waveguide. The sample is located in a microwave cavity between the poles of an electromagnet.

When the sample absorbs there is a reduction in reflected microwaves from the cavity which goes to the detector on the right side of the microwave bridge. When the sample in the microwave cavity is a magnetic material, the absorption of microwave is determined by the ferromagnetic properties and is called ferromagnetic resonance (FMR). FMR employs the similar equipment as EPR.

However, in FMR the rf energy is not absorbed by transitions of the electron spin from the spin down state to the spin up state but rather due to the precession of the total magnetization **M** of the sample about the applied dc magnetic field B_0. The energy is absorbed when the applied frequency is equal to the precessional frequency. This causes the direction of magnetization to flip. The equation of motion for a unit volume of the sample is,

$$d\mathbf{M}/dt = \gamma[\mathbf{M} \times \mathbf{B}_0] \tag{8.5}$$

where γ is the magnetogyric ratio, M the magnetization, B_0 the applied magnetic induction and ω_0 is the precessional frequency.

The resonance frequency of the FMR signal depends on the shape of the sample. For a flat rectangular ferromagnetic plate, such as a thin film, the resonance condition for B_0 perpendicular to the plate is,

$$\omega_0 = \gamma[\mathbf{B}_0 - 4\pi\mathbf{M}] \tag{8.6}$$

For the magnetic field parallel to the surface of a thin film, the resonance frequency is given by,

$$\omega_0 = \gamma(\mathbf{B}_0[\mathbf{B}_0 - 4\pi\mathbf{M}])^{1/2} \tag{8.7}$$

It is seen that ω_0 depends on the orientation of the magnetic field with respect to the geometry of the sample. Different expressions are obtained for spheres and cylinders. The major characteristics that distinguish FMR signals from ESR absorptions are strong temperature dependence of ω_0 because of the dependence of **M** on temperature and temperature dependent line widths which generally increase with lowering temperature.

Another major difference is that FMR spectrum shows a signal centered at zero magnetic field. Figure 8.7 shows the spectrum of

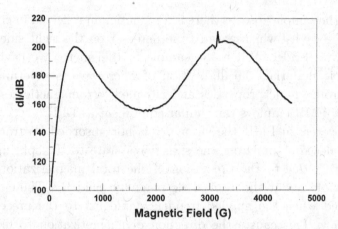

Figure 8.7 Ferromagnetic resonance spectra of nickel at room temperature. (Ref. 1)

nickel particles at room temperature.[1] The spectrum consists of two lines, a broad line in the vicinity of 3000 G and a more intense line centered at zero magnetic field. The broad line at high field is due to the absorption of microwave radiation due to the reorientation of the magnetization vector. The presence of the low field non resonant absorption signal is a well established indication of ferromagnetism in materials. The signal occurs because the permeability in the ferromagnetic state depends on the applied magnetic field increasing at low fields to a maximum and then decreasing. Microwave absorption is proportional to the surface resistance of the sample. Since the surface resistance depends on the square root of the permeability, the microwave absorption depends non-linearly on the strength of the dc magnetic field resulting in a non-resonant derivative signal centered at zero field. This signal is not present in the paramagnetic state and emerges as the temperature is lowered below the Curie temperature, T_c. It provides a unique signature of the presence of ferromagnetism.

8.3 Ferromagnetic Quantum Wells

If one dimension is reduced to the nanometer range and the other two remain large, the structure is called a quantum well. The most obvious example of such a structure is a thin film which has nanometer

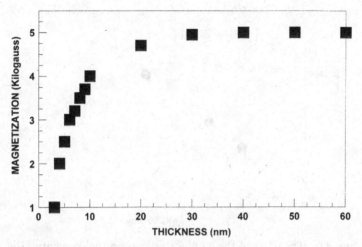

Figure 8.8 Dependence of the saturation magnetization on the thickness of a nickel film on a copper substrate. (Ref. 2)

thickness. Figure 8.8 shows the dependence of the saturation magnetization on the thickness of an electro-deposited nickel film on a copper substrate.[2] Below about 30 nm there is the onset of a substantial reduction in the saturation magnetization. The reason for this decrease is because of the increased fraction of magnetic atoms on the surface, below 40 nm, which lose their ferromagnetic order. Figure 8.9 shows the effect of film thickness on the coercivity.[2] Notice that at a thickness of 2 nm there is no hysteresis indicating the film is superparamagnetic. Ferromagnetic resonance has also been used to study ferromagnetic wells. The studies have been made on ultra thin films of Au/Fe deposited by molecular beam epitaxy on the (001) surface of thin crystals of Gallium Arsenide. The films were 20 atomic layers thick of gold on top of a 10 atomic layers of iron. Figure 8.10 shows the FMR spectra at a number of different temperatures for the dc magnetic field in the plane of the film parallel to the [110] direction of the GaAs substrate.[3] The spectra are asymmetric and shift to lower magnetic fields as the temperature is lowered. The line width also increases with decreasing temperature. These effects are characteristic of FMR spectra.

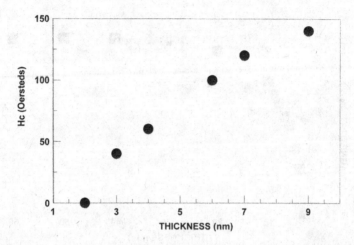

Figure 8.9 The effect of the Ni film thickness on the coercivity. (Ref. 2)

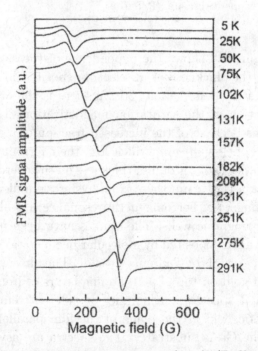

Figure 8.10 Ferromagnetic resonance spectra of an Au/Fe film at a number of different temperatures for the magnetic field in the plane of the film. (Ref. 2)

8.4 Dichalogenides

Layered transition metal dichalcogenides, discussed in Chapter 7, have the potential to enable fabrication of magnetic quantum wells. A typical member of this family of materials is vanadium disulphide (VS_2) which in its bulk form is ferromagnetic. The V^{4+} ion has a large magnetic moment of $3\mu_B$. One crystal structure of this material consists of parallel layers of VS_2 where the S has a charge of -2. The layers are weakly bonded to each other in the lattice by Van der Waals potentials. There are two stable phases for this material designated as the 1T and 2H phases. Examples of the structure of these phases are shown in Figures 7.8 and 7.9. Because of the weak interaction between the VS_2 layers, it has been possible to produce thin nanosheets of VS_2 by exfoliation.[4] While bulk VS_2 is a ferromagnetic metal, evidence for ferromagnetism in a single nanosheet remains to be determined. However, there are theoretical predictions that such sheets should be ferromagnetic.[5]

8.5 Magnetic Quantum Wires

When two dimensions are reduced to nanometer length and one remains large it is a quantum wire. Single walled carbon nanotubes, discussed in Chapter 4, are an example of such a wire. Magnetic wires made of Fe_3O_4 have been fabricated. Changing the diameter can be used to control the magnetic properties of the wires. Figure 8.11 is a plot of the coercivity of the wires as a function of the diameter.[6] Below about 50 nm the coercivity is zero meaning that the magnetization versus magnetic field displays no hysteresis and the wires are superparamagnetic.

Copper oxide (CuO) undergoes a transition from a paramagnetic state to an antiferromagnetic state at 230 K. When CuO is fabricated into nanorods having diameters of 30 to 40 nm and lengths of 100 nm to 200 nm, it displays ferromagnetism. Figure 8.12 is a plot of the measured magnetization versus dc magnetic field at room temperature showing a substantial magnetization.[7] This is a good example of how lowering the dimensions of a material to nanometers can change

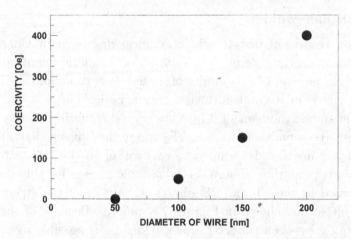

Figure 8.11   Coercevity of Fe_2O_3 wires versus the diameter of the wires. (Ref. 2)

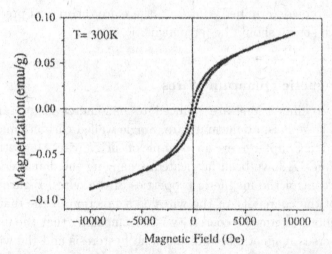

Figure 8.12   Magnetization of nanorods of CuO versus the magnetic field at room temperature. (Ref. 7)

its magnetic properties. In this case making a non magnetic material magnetic.

There are bulk materials which contain magnetic wires as sub units. Some examples are Li_2CuO which has CuO_2 chains. $CsMnCl_3$ $2H_2O$ and $CuCl_2$ are other examples. Figure 7.16 shows the crystal structure of $CuCl_2$ showing the bulk solid contains chains of $CuCl_2$. These chains weakly interact with each other in the lattice and can be approximated as free chains. The structures are important because they allow a test of various theories of magnetism such as the Ising model, for which an exact solution can be obtained only for a one dimensional magnetic system.[8] It has been shown to account fairly well for the temperature dependence of the susceptibility through the paramagnetic to antiferromagnetic transition in $CuCl_2$.[9]

Consider a lattice where each site has a spin S having values ± 1 which can align either parallel or anti-parallel to an applied magnetic field H. In the Ising model the magnetic energy of the lattice is given by,

$$E = J \sum_{i,j} S_i S_j - \mu H \sum_i S_i \qquad (8.8)$$

The first term is summed over the nearest neighbor pairs and the second over all spins of the lattice. For a ferromagnetic material J is positive and for an antiferromagnet it is negative. For a one dimensional system such as $CuCl_2$ in high magnetic fields it has been shown that the Ising model gives the temperature and magnetic field dependence of the susceptibility as,[9]

$$X = [N\mu^2/kT][\exp(-4J/kT) + (\mu H/kT)^2]^{-1/2} \qquad (8.9)$$

This equation has been shown to well describe the temperature dependence of the susceptibility in $CuCl_2$.

A transition from the paramagnetic state to a ferromagnetic or antiferromagnetic state is preceded by fluctuations of the magnetic moments. In effect the randomly oriented magnetic moments have to undergo dynamical reorientation to achieve the aligned orientation in the magnetic phase. The details of the dynamics of the fluctuations depend on the dimensionality of the magnetic system. In one dimensional systems, the fluctuations start at much higher temperatures above the transition temperature than higher dimensional materials. The presence of these fluctuations has been observed by electron paramagnetic resonance measurements. The fluctuations cause temperature dependent deviations of the line width and g values as the transition temperature is approached from above. Further theoretical modeling of the fluctuation process is less complex than in higher dimensions. Thus these one dimensional systems make good materials to test the validity of various proposed models of magnetic fluctuations and this has motivated a number of experimental studies using electron paramagnetic resonance. Generally, the EPR spectrum of a magnetic material is a broad line resulting from the magnetic interaction with the nearest neighbor magnetic atoms. The compound $CsMnCl_3$ $2H_2O$ has a one dimensional linear chain of Cl-Mn^{2+}-Cl-Mn^{2+} which becomes antiferromagnetic at 4.89 K. Starting at about 20 K there is a marked shift in the magnetic field position of the Mn resonance with lowering temperature when the magnetic field is parallel to the chain as shown in Figure 8.13.[10] There is also a broadening of the line as the transition temperature is approached. These effects are due to fluctuations of the magnetic moments prior to the formation of the antiferromagnetic phase. In the context of the one dimensional Heisenberg theoretical representation, models have been developed to predict how the fluctuations produce the

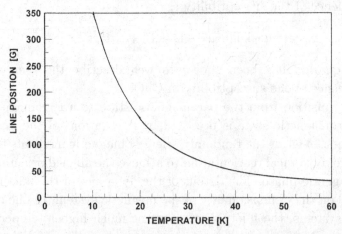

Figure 8.13   Shift of the magnetic field position of the EPR spectrum of Mn^{2+} in the paramagnetic phase of the linear chain material $CsMnCl_2$ $2H_2O$ as the antiferromagnetic state is approached. (Ref. 2)

line broadening and shifts in the field position with temperature. For example the theory predicts the line width should increase as $1/T^{2.5}$, which is in reasonable agreement with experimental measurements.

8.6 Building One Dimensional Magnetic Arrays One Atom at a Time

The scanning electron microscope has been described in Chapter 4 and is illustrated in Figure 4.7. It has been used to build nanosized low dimensional magnetic structures atom by atom on the surface of materials.[11] An adsorbed atom is held on the surface by chemical bonds with the atoms of the surface. When such an atom is imaged in an STM, the tip has a trajectory of the type shown in Figure 8.14a. The separation between the tip and the adsorbed atom is such that any forces between them are small compared to the forces binding the atom to the surface, and the adsorbed atom will not be disturbed by the passage of the tip over it. If the tip is moved closer to the adsorbed atom, as shown in Figure 8.14b such that the interaction of the tip and the atom is greater than that between the atom and the surface, then the atom can be dragged along by the tip. At any point in the scan the atom can be re-attached to the surface by increasing the separation between the tip and the surface. In this way adsorbed atoms can be re-arranged on the surfaces of materials, and structures can be built on the surfaces atom by atom. The surface of the material has to be cooled to liquid helium temperatures in order to reduce thermal vibrations, which may cause the atoms to undergo thermally induced diffusion, thereby disturbing the arrangement of atoms being assembled. Thermal diffusion is a problem because this method of construction can only be carried out on materials in which the lateral or in-plane interaction of the adsorbed atom and the atoms of the surface is not too large. The manipulation also has to be done in an ultra high vacuum in order to keep the surface of the material clean. The method has been used to make low dimensional arrays of iron atoms on the (111) surface of ultra clean copper. Figure 8.15 shows a STM image of a one dimensional chain of seven iron atoms made by this process.[11] The linear chain has anitferromagnetic order.

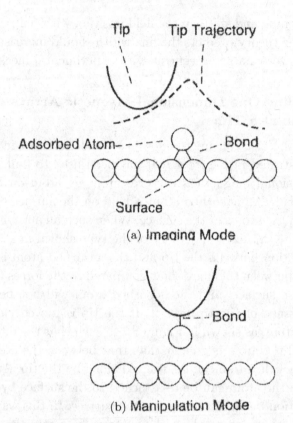

Figure 8.14   Illustration of how the STM is used to build atomic structures on the surfaces of clean metals such as copper.

The dark images are the iron atoms with spin down and the lighter images are iron atoms with spin up. It was observed that when the chain contained an even number of iron atoms there is no antiferromagnetic order. For example a chain consisting of 6 iron atoms had four atoms in the spin down orientation and two in the spin up orientation. These structures can provide an experimental test of the Ising model mentioned previously as it has an exact solution for a one dimensional magnetic system. It was found that there were deviations of the magnetic field dependence of the magnetization from the predictions of the Ising model. The reason for this is not understood and presently under investigation.

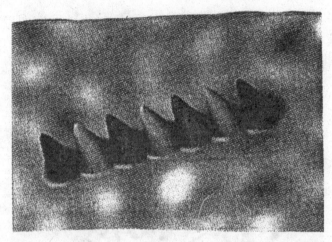

Figure 8.15   STM image of a linear chain of iron atoms built on the surface of copper using the STM. (Ref. 2)

8.7 Ferromagnetism in Carbon and Boron Nitride Nanotubes

While there are many theoretical predictions of the possibility of ferromagnetism in single walled carbon nanotubes (SWNTS) having vacancies or doped with various atoms such as boron or nitrogen, there are no substantiated reports of intrinsic ferromagnetism in SWNTS. One of the problems is the presence of the magnetic catalysts necessary for synthesis would mask any intrinsic ferromagnetism. Multi-walled carbon nanotubes grown inside alumina templates were shown to contain many defects. When subjected to hydrogen gas the tubes adsorbed considerable hydrogen and were shown to be ferromagnetic having a Curie temperature near 1000 K.[12] In another experiment it was shown that multi-walled carbon nanotubes can become magnetized when they are in contact with a magnetic material.[13] The magnetism was observed using a magnetic force microscope. Ferromagnetism above room temperature has been observed in double walled carbon nanotubes which were subjected to an acid treatment to remove the iron catalyst. The magnetization at room temperature was in the order of 2.2 emu/g which was considerably greater than the estimated value of the magnetization from the

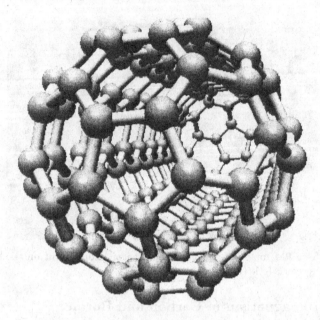

Figure 8.16　Illustration of a capped carbon nanotube.

residual iron catalyst which was determined to be 0.022 emu/g. The ferromagnetism was attributed to adsorbed hydrogen on the tubes due to the acid purification process.

Capped nanotubes, shown in Figure 8.16, which do not require metal catalysts for synthesis, show evidence for ferromagnetism when exposed to oxygen at high pressures.[14] Because the tubes were closed at both ends they were heated to between 400°C to 500°C which opens the ends of the tubes allowing oxygen to penetrate the interior of the tubes. Magnetism was observed at low temperature, in the vicinity of 50 K, in the treated material. The magnetism is attributed to the presence of solid oxygen inside the tubes or adsorbed on the surface of the tubes. There have been no experimental reports of ferromagnetism in boron nitride or boron nanotubes.

While there have been only a few experimental hints of magnetism in SWNTS, there have been many theoretical predictions of the possibility of magnetism induced by doping, side wall functionalization or defects.

8.8 Magnetism in Graphene

Ferromagnetism has been observed in a two dimensional carbon nanosheet. similar to that illustrated in Figure 6.1.[15] The sheets were fabricated using an inductively coupled radio frequency technique. A chamber containing an RF source is used to generate a plasma from a mixture of argon and methane gas. The graphite layers are formed on a silicon substrate from carbon atoms which are products of the decomposition of methane, CH_4. The substrate was held at a constant temperature of 400°C. Various spectroscopic techniques such as atomic force microscopy, x-ray diffraction and Raman spectroscopy were used to verify the existence of the one dimensional graphite layer. Only the material made at the longest deposition time of 120 minutes shows evidence of ferromagnetism. Raman spectroscopy was used to characterize the sample made at different deposition times. The spectra consists of two strong lines at 1588 cm^{-1} (the G mode) and 1322 cm^{-1} (the D mode) which have been discussed in Chapter 6. The material made with the two lowest deposition times had a I_g/I_d ratio close to one indicating a highly defected structure and low degree of graphene formation. The material with the longest deposition time has a ratio of 2.7 indicating a low defect content.

There have been other observations of ferromagnetism in graphene like structures. For example graphene made from graphene oxide which can be chemically synthesized has been shown to display ferromagnetism having a maximum magnetization of 0.010 emu/gram.[16] The origin of the ferromagnetism has not been determined but is likely due to the presence of some kind of defects. One possibility is that the hydrogen released in the decomposition of methane bonds to the surface of the graphene plane. There have been theoretical predictions that surface hydrogenated graphene could be ferromagnetic.

One defect that is not likely to cause the ferromagnetism is a carbon vacancy. Carbon vacancies have been produced in graphene by proton irradiation and the resulting material has been shown to be paramagnetic with the spin 1/2 entities localized around the vacancy.[17] However, it was not possible to achieve more than one magnetic moment per 1000 carbon atoms. Graphene with a greater vacancy concentration would likely be unstable. DFT calculations of the minimum energy structure of graphene ribbons having a single vacancy indicate the ribbons are distorted from planarity.[18] Since ferromagnetism results from exchange interaction between unpaired spins, which requires that the distance between the spins be relatively small, it is unlikely that carbon vacancies in the graphene plane can be the cause of the ferromagnetism. It should be emphasized that the observed magnetizations in all observations are quite small and most of the observations have not been replicated to date. There is much more research needed to establish the existence of ferromagnetism in graphene and determine its origin.

The energy levels of a solid depend on the K vector, which in the free electron model of metals is given by equation (1.7) where $K = n/2L$ for the one dimensional lattice. In order to fully characterize the energy levels of a solid their dependence on the K vector is needed. The tight binding model discussed in Chapter 2, can be used to calculate the K dependence of the energies of graphene.[19] The calculations reveal some unusual features, which bear on the possibility of ferromagnetism in graphene. Figure 8.17b and 8.17a show the calculated dependence of the energy of the highest occupied orbital (HOMO) and the lowest unoccupied orbital (LUMO) for a zigzag

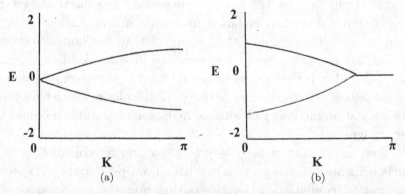

Figure 8.17 Calculation of the dependence of the highest occupied orbital and the lowest occupied orbital versus K vector for a zigzag graphene ribbon (b) and an armchair graphene ribbon (a). (Ref. 19)

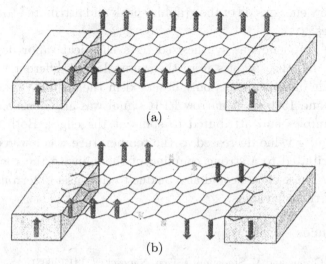

(a)

(b)

Figure 8.18   A magnetoresistive device based on zigzag nanoribbons having edge states.

and armchair ribbon having a width of 4 carbon rings. The energies are normalized to t, which is the nearest neighbor overlap integral in the tight binding model. The results for the zigzag ribbon show an unusual coincidence of the energy levels of the HOMO and the LUMO near K = π. This degeneracy is not predicted for graphite or the armchair ribbons. It is found that in this region where the energies are coincident charge is localized on the zigzag edges. These states are referred to as edge states. Theory also predicts that there are magnetic moments associated with these edge states and that these moments are ordered ferromagnetically on one side of the ribbon and anitiferromagnetically on the opposite side.

There is some preliminary experimental evidence for the edge states in graphene ribbons. Near edge x-ray absorption fine structure spectroscopy (NEXAFS) has been used to investigate this issue.[20] In this experiment an x-ray photon excites an electron from the core carbon 1s level, producing a photo-electron emission.The energy of the emitted electron is measured. In graphite there is a peak at 285.5 eV. corresponding to a transition from the carbon 1s level to the LUMO state. In graphene an additional small peak was observed on the low energy side of the graphite peak and attributed to spins at the edges.

The graphene was synthesized by a chemical vapor deposition method and the samples were then annealed at different temperatures. The new peak was only observed in the samples annealed at 1000°C and 1500°C. A narrow EPR signal was also observed in the same samples and attributed to spins at the edges. Both the line width and g value decreased as the temperature was lowered which was attributed to a strong coupling of the spins to the conduction carriers. These effects could also be due to the onset of ferromagnetic order of the edge spins.

References

1. F. J. Owens and V. Stepanov, J. Exp. Nanosci. *3*, 141 (2008)
2. *Physics of Magnetic Nanostructures,* p58, 59, 60, 61, 64, 66, 91 by Frank J. Owens, John Wiley & Sons, Hoboken, NJ, 2015
3. *Magnetic Nanostructures,* p66, Edited by B. Aktins, L. Tagirov and F. Miknlov, Springer Verlag, N.Y.,2007
4. J. Feng *et al.* J. Am. Chem. Soc. *133*, 17832 (2011)
5. Y. Ma *et al.* ACSNano, *6*, 1695 (2012)
6. L. Zhang and Y. Zhang, J. Magn. Magn. Mater. *321*, L15 (2009)
7. H. Xiao *et al.* Solid State Commun. *141*, 431 (2007)
8. *Principles of the Theory of Solids* by J. M. Ziman, page 299, Cambridge University press, 1964
9. C. G. Barralough and C. F. No, Trans. Faraday Soc. *60*, 836 (1964)
10. F.J. Owens, Phase Transitions 5, 842(1985)
11. A. A. Khajetoodians *et al.* Nature Physics, *8*, 497 (2012)
12. A. L. Frieman, Phys. Rev. *B81*, 115461 (2010)
13. O. Cespedes, J. Phys. Condensed Matter, *16*, L155 (2004)
14. S. Bandow, T. Yamaguchi and S. Iijima, Chem. Phys.Lett. *401*, 380 (2005)
15. B. P. C. Rao *et al.* Phil. Mag. *90*, 3463 (2001)
16. Y. Wang *et al.* Nanoletters, *9*, 220 (2009)
17. R. R. Nair *et al.* Nature Physics, *8*,199 (2012)
18. M. Miller and F. J. Owens, Chem. Phys. Lett. *570*, 42 (2013)
19. K. Nakada, M. Fujita, G. Dresselhaus and M. S. Dresselhaus Phys. Rev. *B54*, 1795 (1996)
20. V. L. Joly *et al.* Phys. Rev. *B81*, 245428 (2010)

Chapter 9

Superconductivity in Low Dimensional Materials

9.1 Properties of the Superconducting State

In a superconductor there is temperature, T_c, where the resistance to direct current and low frequency ac current becomes zero. Figure 9.1 is a plot of the dc resistance normalized to the room temperature value versus temperature for the superconductor $Hg_{0.8}Pb_{0.2}Ba_2$ $Ca_2Cu_3O_{8+x}$ which reaches zero resistance at 130 K, the highest temperature of any superconductor at ambient pressure.[1]

The carriers of current in the superconducting state have a charge 2e, twice the electron charge. This means that the electrons at the Fermi level of the metal, which carry the current, are bound in pairs which are called Cooper pairs. The existence of these bound pairs alters the energy band gap picture of the metal in the superconducting state. In the superconducting state the observation of bound electron pairs implies that there must be an energy gap, Δ, at the Fermi level. The magnitude of this superconducting gap corresponds to the binding energy of the electron pairs. It is the energy difference between the normal electrons and the bound electron pairs at the Fermi level.

One of the major challanges in the development of understanding of superconductivity was to explain how two negatively charged

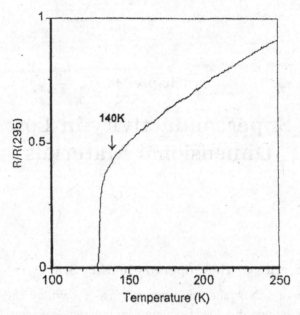

Figure 9.1   Temperature dependence of the resistance of the $Hg_{0.8}Pb_{0.2}Ba_2$ $Ca_2Cu_3O_{8+x}$ superconductor normalized to its value at 295 K. (Ref. 1)

electrons can be bound into pairs despite the repulsive electrostatic force between them.

A critical experiment, showed that in isotopically labled mercury, the transition temperature was shifted proportional to $1/m^{1/2}$, where m is the mass of the Hg isotope. This, indicated that lattice phonons were involved in the binding of the Cooper pairs.[2] This was explained in the BCS theory of superconductivity.[3]

A classical (non quantum mechanical) description can be used to give some insight into how electrons can be bound into pairs in a solid. Because the valence electrons are detached from the atoms and move freely through the lattice, the atoms of the metal have acquired a positive charge. When the conduction electrons move past these positively charged atoms, the atoms are attracted to the electrons and there is a region of local distortion in the lattice. This distorted region is slightly more positively charged than the rest of the lattice and it follows the electron as it moves through the lattice. This more positive region may attract a distant electron and cause it to follow the distortion as it moves through the lattice, in effect forming a bound electron pair. The binding energy of the two electrons is in the order of 10^{-4} eV and the separation of the electrons is about of 10^3 angstroms, about 300 lattice spaces for the metallic low temperature superconductors. Thus the quantum mechanical wavelength of the Cooper pairs is much larger than the diameters and spacings of the atoms of the solid. So the Cooper pairs do not "see" the atoms of the lattice and are not scattered by them. Thus the material has zero resistance. The spins of the electrons of the pair are oppositely aligned so a bound Cooper pair has zero spin and is a Boson. This means that at absolute zero all Cooper pairs will be in the ground state and have the same energy and therefore the same wavelength. Thus not only is the wavelength of the pairs very large but all the pairs have the same wavelength. Further it turns out that the phase of the wave of every pair is the same as that of any other pair. The Cooper pairs have phase coherence analogous to the waves of light produced by a laser. In other words the motion of the pairs in the lattice is correlated. It is this remarkable property of the quantum mechanical wave describing the Copper pairs that accounts for their movement through the lattice without scattering and the resulting zero resistance of the superconducting state.

The second major characteristic of the superconducting state is called the Meissner Effect.[4] If a material which is superconducting is cooled below its transition temperature in a magnetic field, H, the magnetic flux density in the bulk of the material will be expelled at Tc. This behavior is most commonly observed by measuring the temperature dependence of the magnetization M or susceptibility χ = M/H of the sample. Since the flux density B inside a superconductor is related to the applied field H by,

$$B = \mu_0 H + M = \mu_0 H(1 + \chi) \qquad (9.1)$$

B will be zero inside the superconductor and in the MKS system used in writing equation (9.1), $\chi = -1$. The material in effect behaves like a perfect diamagnet. Figure 9.2 shows the results of a measurement of

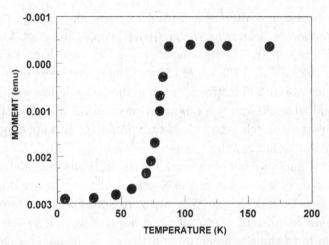

Figure 9.2   Temperature dependence of the magnetization of the $Hg_{0.8}Pb_{0.2}Ba_2$ $Ca_2Cu_3O_{8+x}$ superconductor. (Ref. 1)

the temperature dependence of the magnetization for a single crystal of $Hg_{0.8}Pb_{0.2}Ba_2$ $Ca_2Cu_3O_{8+x}$.[1] Although magnetic flux is excluded from the bulk, it penetrates the surface layers of the superconductor. The temperature dependence of the penetration depth λ of the magnetic flux has been shown to obey the equation,[5]

$$\lambda(T) = \lambda_o/(1 - [T/T_c]^4)^{1/2} \tag{9.2}$$

where λ_o is the penetration depth at absolute zero. Another parameter that characterizes the superconducting state is the coherence length, ξ, which is a measure of the distance between the electrons or holes in the Cooper pairs. These parameters are in the order of nanometers for most superconductors.

Two types of superconductors can be identified by their behavior in a dc magnetic field. In a type I superconductor, as the external magnetic flux density, B_{appl}, is increased it does not penetrate the bulk of the material in the superconducting state until a field is reached called the critical field, B_c, above which the superconducting state no longer exists. In a type II superconductor no magnetic field penetrates the bulk of the sample until a field B_{c1} is reached referred to as the lower critical field. At B_{c1} flux density begins to penetrate the sample but does not remove all of the superconducting state. The flux penetrates the sample in tube like normal regions and can only have a value in these regions of $\varphi_0 = h/2e = 2.07 \times 10^{-15}$ Wb where h is Planck's constant. As the field is further increased to a value B_{c2}, referred to as the upper critical field, flux density continues to penetrate the sample. At B_{c2} the superconducting state is totally removed.

The existence of a critical magnetic field which removes the superconducting state implies there is an upper limit to the current density that can be carried by the superconductor because current produces a magnetic field.

9.2 Copper Oxide Superconductors

The copper oxide superconductors were first discovered in 1986.[6] The normal conductivity and superconductivity largely occurs in two dimensional planes for all of the different materials. Figure 9.3 illustrates the structure of $Hg_{0.8}Pb_{0.2}Ba_2 Ca_2Cu_3O_{8+x}$ which reaches zero resistance at 130 K representing a typical illustration of the structure of the copper oxide superconductors. It consists of parallel planes of copper and oxygen. It is believed this two dimensional property plays an important role in causing the higher transition temperatures. The conduction in the normal state is by hole hopping and is largely confined to these planes. The superconducting transition temperature depends on the hole concentration which is determined by the oxygen concentration.

The fact that the current is largely confined to the copper oxide planes means that the band structure of the material can be approximated by a two dimensional model. In effect the copper oxide superconductors are approximately two dimensional superconductors. Figure 9.4 shows the energy versus density of states for a simple two dimensional array of copper and oxygen atoms where each copper atom is bonded to four nearest neighbor oxygen atoms in a square array. The top occupied band is formed from the highest occupied orbitals of the atoms (HOMO) and the first unoccupied band (LUMO) is formed from the first empty orbitals of the atoms.

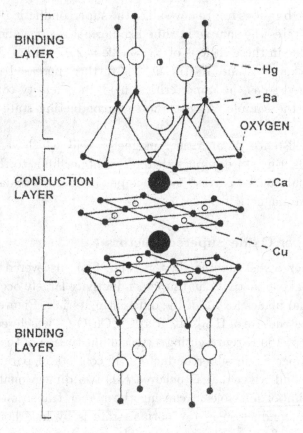

BINDING
LAYER

CONDUCTION
LAYER

BINDING
LAYER

Hg

Ba

OXYGEN

Ca

Cu

Figure 9.3   Crystal structure of Hg-Ba-Ca-Cu-O superconductor.

In this simple model the energies of the filled lower band $E^H(K)$ and the empty upper band $E^L(K)$ are,

$$E^L(K) = E^L + 2t^L(\cos K_x + \cos K_y) \qquad (9.3)$$

$$E^H(K) = E^H + 2t^H(\cos K_x + \cos K_y) \qquad (9.4)$$

E^L and E^H are the energies of the LUMO and HOMO free atom orbitals and t^H and t^L are the corresponding overlap integrals. The peaks in the density of states near the middle of each band are unique to two dimensional systems and are the van Hove singularities discussed earlier. They can contribute to increasing the transition

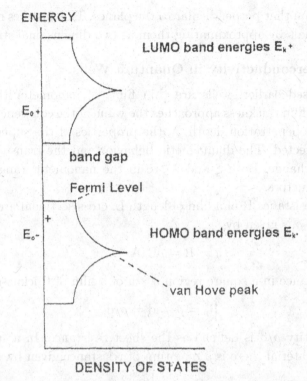

Figure 9.4   The energy versus density of states for a simple two dimensional array of copper and oxygen atoms where each copper atom is bonded to four nearest neighbor oxygen atoms.

temperature. They are a result of the many combinations of K_x and K_y giving the same energy. Because of the reduced dimensionality, the band gap is anisotropic having values approximately $1.75kT_c$ parallel to the c axis and $2.7kT_c$ parallel to the copper oxide planes. The separation between the holes in the Cooper pairs in the copper oxide planes is typically in the order of 20 Å, significantly smaller than in the low temperature metallic superconductors where this distance can be as large as 1000 Å. Another consequence of the two dimensional nature of the copper oxides is anisotropy of the electrical transport both in the superconducting and the normal state. The conductivity parallel to the planes is about 2 to 4 orders of magnitude greater than that perpendicular to the planes. This justifies modeling the materials by approximating them as two dimensional structures.

9.3 Superconductivity in Quantum Wells

As discussed earlier, wells are thin films of nanometer thickness. When the film thickness approaches the value of the coherence length, .ξ, and the penetration depth, λ, the properties of the superconductor are affected. The diamagnetic behavior and the transition temperature change. Both ξ and λ are in the nanometer range for all superconductors.

The resistance R of a film of length L, cross sectional area A and resistivity ρ is given by,

$$R = \rho L/A \tag{9.5}$$

The resistance in a square region a $\times$ a of a film of thickness d is

$$R_s = \rho a/ad = \rho/d \tag{9.6}$$

The quantity ρ/d is defined as the sheet resistance. In a superconducting material there is a quantum of resistance given by,

$$R_Q = h/(2e)^2 \tag{9.7}$$

where 2e is the charge of the Cooper pairs. Inorder for a well to superconduct, the well must be thick enough so that R_s above T_c is less than R_Q. If R_s is slightly less than R_Q, the transition temperature will depend on R_s as shown in Figure 9.5 for a quantum well of Pb.[7] Figure 9.6 shows the dependence of T_c on inverse of the well thickness.[7]

9.4 The Proximity Effect

When films of two different superconductors are in direct contact with each other, the wave function of the Cooper pairs of each penetrate into the other superconductor. This affects the transition temperature of the two materials. Generally the transition temperature will shift to some value between that of each superconductor. When a non superconducting metal film such as Cu is in contact

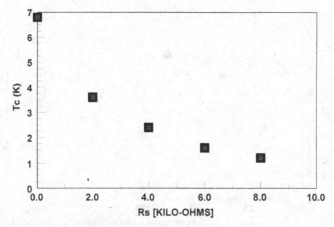

Figure 9.5 Dependence of the transition temperature of a Pb film on the sheet resistance. (adapted from Ref. 7)

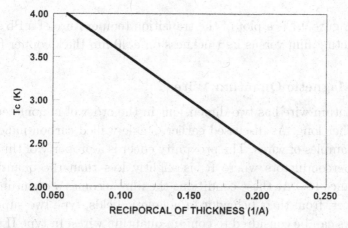

Figure 9.6 Dependence of the transition temperature of a Pb film on the inverse of the film thickness. (adapted from Ref. 7)

with superconducting film such as Pb, Cooper pairs will cross the interface into the non superconducting metal. This will cause a reduction in the superconducting transition temperature. These anomalies are referred to as the Proximity effect. The transition temperature depends on the thickness of both superconducting film and metal

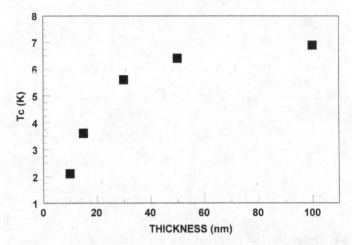

Figure 9.7   Plot of the transition temperature of a Pb superconducting film versus its thickness on a 40 nm thick copper film. (adapted from Ref. 8)

film. Figure 9.7 is a plot of the transition temperature of a Pb superconducting film versus its thickness on a 40 nm thick copper film.[8]

9.5 Magnetic Quantum Wires

A quantum wire has two dimensions in the order of nanometers and the other long. As discussed earlier, single walled carbon nanotubes are examples of wires. The proximity effect is a property of thin films of superconductors where R_s is slightly less than the quantum of resistance R_Q. At these conditions the film becomes a quantum well. However, from their behavior in magnetic fields, type two superconductors can be considred to contain quantum wires. In type II superconductors when a magnetic field having a value greater than B_{C1} is applied there is a partial penetration of the magnetic field into the superconductor. The magnetic flux penetrates into the superconductor in the form of tubes of normal regions called vortices. Each vortex encloses one quantum of flux, φ_0, having a value h/2e. It has a radius of, ξ, which can be 4 nm or less depending on the superconductor. Thus from the electrical perspective thin films of superconductors under certain conditions can be classified as quantum wells. From the magnetic behavior type II superconductors can be classified as quantum wires.

9.6 Iron Selenide

In 2008 superconductivity was discovered in iron selenide, FeSe, having a T_c of 8 K[9]. Untill the observation of superconductivity in $RuSr_2Gd_{0.5}Eu_{0.5}Cu_2O_8$, it was believed that superconductivity could not co exist in a material having magnetic atoms. FeSe is a layered material analogous to the dichalognides discussed earlier. Its structure is illustrated in Figure 9.8. Subsequently superconductivity was discovered in a number of other layered iron materials. Such as $FeSe_{0.5}Te_{0.5}$, $K_{0.82}Fe_{1.63}Se_2$, $Ba_{0.48}Fe_{2.19}Se_{1.16}$ and $K_{0.87}Fe_{2.19}Se_2$.[10-13]

The intriguing aspect of these results is the observation that as the separation between the layers increased T_c increased. Figure 9.9 is a plot of the T_c versus the distance between the layers of the materials listed above.[12] This result suggests the possibility that single layer of FeSe could have a high transition temperature. Researchers focused on this possibility. A two layered FeSe structure, grown on the (001) surface of $SrTiO_3$ was shown to have an onset temperature

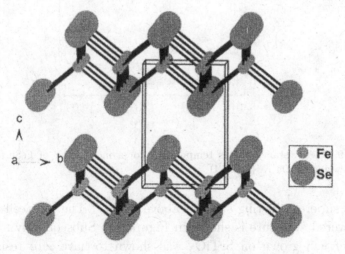

Figure 9.8   Structure of the unit cell of FeSe.

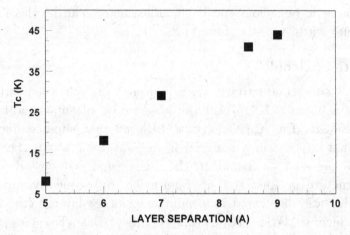

Figure 9.9   A plot of T_c versus the distance between the layers of various Iron Selenide superconductors. (adaped from Ref. 12)

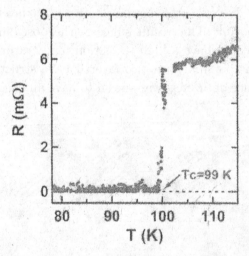

Figure 9.10   Resistance versus temperature for a mono layer of FeSe showing zero resistance at 99 K. (Ref. 14)

to the superconducting state of above 50 K.[13] The unit cell of the two layered structure is shown in Figure 9.8. Subsequently a single layer of FeS grown on $SrTiO_3$ was shown to have zero resistance at 99 K.[14] Figure 9.10 shows the resistance versus temperature for this two dimensional material. This represents the first observation of superconductivity in a truly two dimensional material and opens the possibility that superconductivity could exist in other two dimensional solids perhaps having even higher transition temperatures.

9.7 Superconductivity in Single Walled Carbon Nantoubes

The initial reports of superconductivity in SWNTs were received with some skeptism.[15-17] Firstly, because carbon materials were not know to be superconductors and because superconductivity should be quenched in SWNTs because of long wavelength thermal fluctations know to exist in one dimensional materials.

In the first observation of superconductivity in SWNTs, the tubes were fabricated in the empty linear channels of AlPO4-5 (AFI) zeolite crystals. Tripropylamine (TPA) was incorporated into the linear channels of AFI. The AFI containing the TPA was heated in a vacuum chamber to 580°C for several hours. The carbon atoms resulting from the decomposition of TPA formed SWNTs. High Resolution transmission electron microscopy and x-ray diffraction showed that the tubes had a diameter of 4 angstroms, the smallest diameter tubes ever synthetized. The normalized temperature dependence of the susceptibility indicated the onset of superconductivity at 10 K.[15] Subsequently, in later work the temperature dependence of the resistivity was reported and it indicted the onset of superconductivity in approximately the same temperature range.[18] Measurements of the radial breathing modes by Raman spectroscopy indicated there were three kinds of SWNTS in the zeolite matrix, a zig zag (5,0), an armchair (3,3) and a chiral (4,2).[19] The SWNT or SWNTS responsible for the superconductivity were not determined. Because the temperature dependence of the specific heat could be accounted for by the Ginzburg- Landu phemonological model, it is likely that the superconductivity can be accounted for by the BCS theory.[20] This indictes the SWNT would have to be metallic. Theoretical calculations suggested that (5,0) SWNTs were responsible for the superconductivity because these tubes were metallic and had a high electron density of states at the Fermi level which is a prerequisite for superconductivity.[21]

9.8 Superconductivity in Graphene

As discussed in Chapter 6 graphene is a two dimensional material where the carbon atoms have the same arrangement as the carbons in the planes of graphite. It has been found that incorporating Ca atoms between two planes of graphene forming graphene laminates results in superconductivity at 1.8 K.[22]

Superconductivity at 7.4 K has been observed in a few layers of graphene intercalated with lithium.[23] Figure 9.11 shows a measurement of the susceptibility versus temperature in different dc magnetic fields showing the presence of superconductivity.

Angle resolved photo emission spectroscopy shows that when lithium is deposited on a monolayer of graphene, the phonon density of states is significantly modified, causing a significant enhancement of the electron-phonon interaction. An energy gap of one mille eV was observed at the Fermi level indicating the possibility of superconductivity at 7 K.[24] However, to date there have been no measurements of the temperature dependence of the resistance or susceptibility to verify this possibility.

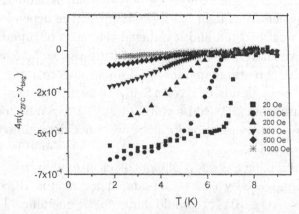

Figure 9.11 The dependence of the susceptibility on temperature for various applied dc magnetic fields for lithium doped graphene. (Ref. 23)

References

1. Z. Iqbal *et al.* Phys. Rev. *B49*, 12322 (1994)
2. C. A. Reynolds, B. Serin, W. H. Wright and L. B. Nesbit, Phys. Rev. *78*, 487 (1950)
3. J. Bardeen, M. L. Cooper and J. Schrieffer, Phys. Rev. *108*, 1175 (1957)
4. W. Meissner and R. Ochsenfeld, Naturwissenschaften. *21*, 787 (1933)
5. F. London and H. London, Proc. Roy. (London) *A149*, 72, 1935
6. J. G. Bednorz and K. A. Muller, Z. Phys. *B64*, 189 (1986)
7. D. B. Haviland, Y. Liu and A. M. Goldman, Phys. Rev. Lett. *62*, 2180 (1989)
8. N. R. Werthamer, Phys. Rev. *132*, 2440 (1963)
9. F. C. Hsu *et al.* Proc. Natl. Acad. Sci. U.S.A. *105*, 14262 (2008)
10. K. W. Yeh, Europhysics Lett. *84*, 37002 (2008)
11. D. Wang *et al.* Phys. Rev. *B83*, 132502 (2011)
12. A. Zhang *et al.* Sci. Rep. *3*, 1216 (2013)
13. Q. Y. Wang, Chin. Phys. Lett. *29*, 037402 (2012)
14. J. Ge *et al.* arXiv 14063435
15. Z. K. Tang *et al.* SCIENCE, *292*, 2492 (2001)
16. M. Kociak *et al.* Phys. Rev. Lett. *86*, 2416 (2010)
17. I. Takesue, Phys. Rev. Lett. *96*, 057001 (2006)
18. Z. Wang *et al.* Phys. Rev. *B81*, 174530 (2010)
19. J. T. Ye and Z. K. Tang, Phys. Rev. *B72*, 045414 (2005)
20. C. R. Lortz *et al.* Proc. Natl. Acad. Sci. USA, *106*, 7299 (2009)
21. H. J. Liu and C. T. Chan, Phys. Rev. *B66*, 115416 (2002)
22. J. Chapman *et al.* arXiv 1508.00363
23. A. P. Tiwari *et al.* arXiv 1508.06360
24. B. M. Ludbrook, arXiv 1508.05925v2.

Index

Printed in the United States
By Bookmasters